Methods for the Quantitative Assessment of Channel Processes in Torrents (Steep Streams)

T0239591

IAHR Monograph

Series editor

Peter A. Davies
Department of Civil Engineering,
The University of Dundee,
Dundee,
United Kingdom

The International Association for Hydro-Environment Engineering and Research (IAHR), founded in 1935, is a worldwide independent organisation of engineers and water specialists working in fields related to hydraulics and its practical application. Activities range from river and maritime hydraulics to water resources development and eco-hydraulics, through to ice engineering, hydroinformatics and continuing education and training. IAHR stimulates and promotes both research and its application, and, by doing so, strives to contribute to sustainable development, the optimisation of world water resources management and industrial flow processes. IAHR accomplishes its goals by a wide variety of member activities including: the establishment of working groups, congresses, specialty conferences, workshops, short courses; the commissioning and publication of journals, monographs and edited conference proceedings; involvement in international programmes such as UNESCO, WMO, IDNDR, GWP, ICSU, The World Water Forum; and by co-operation with other water-related (inter)national organisations. www.iahr.org

Supported by
Spain Water
and IWHR, China

Methods for the Quantitative Assessment of Channel Processes in Torrents (Steep Streams)

Dieter Rickenmann

Research Unit Mountain Hydrology and Torrents, Swiss Federal Research Institute WSL, Birmensdorf, Switzerland

CRC Press
Taylor & Francis Group
Boca Raton London New York

CRC Press is an imprint of the
Taylor & Francis Group, an **informa** business

A BALKEMA BOOK

Published by:
CRC Press/Balkema
P.O. Box 447, 2300 AK Leiden, The Netherlands
e-mail: Pub.NL@taylorandfrancis.com
www.crcpress.com – www.taylorandfrancis.com

First issued in paperback 2020

ISBN 13: 978-0-367-57495-6 (pbk)
ISBN 13: 978-1-138-02961-3 (hbk)

Visit the Taylor & Francis Web site at
http://www.taylorandfrancis.com

and the CRC Press Web site at
http://www.crcpress.com

Typeset by V Publishing Solutions Pvt Ltd., Chennai, India

Library of Congress Cataloging-in-Publication Data

About the IAHR Book Series

An important function of any large international organisation representing the research, educational and practical components of its wide and varied membership is to disseminate the best elements of its discipline through learned works, specialised research publications and timely reviews. IAHR is particularly well-served in this regard by its flagship journals and by the extensive and wide body of substantive historical and reflective books that have been published through its auspices over the years. The IAHR Book Series is an initiative of IAHR, in partnership with CRC Press/ Balkema – Taylor & Francis Group, aimed at presenting the state-of-the-art in themes relating to all areas of hydro-environment engineering and research.

The Book Series will assist researchers and professionals working in research and practice by bridging the knowledge gap and by improving knowledge transfer among groups involved in research, education and development. This Book Series includes Design Manuals and Monographs. The Design Manuals contain practical works, theory applied to practice based on multi-authors' work; the Monographs cover reference works, theoretical and state of the art works.

The first and one of the most successful IAHR publications was the influential book "*Turbulence Models and their Application in Hydraulics*" byW. Rodi, first published in 1984 by Balkema. I. Nezu's book "*Turbulence in Open Channel Flows*", also published by Balkema (in 1993), had an important impact on the field and, during the period 2000–2010, further authoritative texts (published directly by IAHR) included *Fluvial Hydraulics* by S. Yalin and A. Da Silva and *Hydraulicians in Europe* by W. Hager. All of these publications continue to strengthen the reach of IAHR and to serve as important intellectual reference points for the Association.

Since 2011, the Book Series is once again a partnership between CRC Press/ Balkema – Taylor & Francis Group and the Technical Committees of IAHR and I look forward to helping bring to the global hydro-environment engineering and research community an exciting set of reference books that showcase the expertise within IAHR.

Peter A. Davies
University of Dundee, UK
(Series Editor)

Table of contents

Abstract

An important part of the risk management of natural hazards in mountain regions concerns the hazard assessment and the planning of protection measures in steep headwater catch-ments, i.e. torrent control and slope stabilization. This publication presents an overview of methods to quantify channel processes in steep catchments. The understanding and the quantitative description of channel processes provides an essential basis for the planning of protection measures. In the European Alps, channel processes are mostly triggered by rainfall events and associated runoff processes. Apart from possible flood hazards during an intense rainstorm event, a lot of damage is often caused by fine and coarse sediment which is entrained either in the form of fluvial bedload transport or of a debris flow. Typically, the damage increases with the total amount of sediment transported to the fan during an event, particularly if the water and the sediments leave the channel on the fan. This document mainly discusses the topics of flow resistance, bedload transport, debris flows and the relation between magnitude and frequency of torrential sediment events.

A first focus is put on the calculation of flow resistance in steep channels. Flow resistance is shown to increase considerably in steep channels which are often characterized by very irregular bed morphology and large-sized sediment particles such as boulders and pebbles. Together with limited runoff in small catchments, this produces relatively shallow flow depths. For these conditions, some flow resistance approaches used for flatter streams and rivers are not valid. A recently-developed, new flow resistance law is presented, and a quantitative procedure is introduced which allows to account for high flow resistance in bedload transport calculations.

A second key aspect concerns fluvial bedload transport in steep streams. Steep torrent channels show differences from flatter mountain rivers. Grain size analysis is a prerequisite for the calculation of bedload transport. Several formulae are introduced which may be used for the prediction of bedload transport for a given hydrograph. The quantification of three main elements is discussed, namely initiation of particle motion, transport rate, and accounting for high flow resistance. A serious complication during a flood event may be the entrainment of large woody debris, which may lead to clogging at critical channel locations. Erosion and aggradation of sediment may also become a crucial process during a flood event.

As a third core area, debris flows and important elements for its hazard assessment are pre-sented. The occurrence of debris flows is discussed in terms of the primary mechanisms and of triggering rainfall conditions. Empirical and semi-empirical equations are introduced to es-timate the main parameters characterizing the flow and deposition behavior of debris flows. Simulation tools are presented, which may be used primarily to estimate the potentially-affected areas on the fan as well as the flow dynamics. A geomorphic assessment of the natural fan surface can provide indications about the process behavior including, for example, the runout distance of former events.

The last focus is on the magnitude and frequency of torrential sediment events. Apart from historic documents, a field-based geomorphic assessment is recommended to arrive at a good estimate of a future event magnitude. A recently-developed procedure is introduced which combines a field assessment with a GIS-based analysis of other factors that may be relevant for sediment supply to channel system and for sediment entrainment along the channels during a rainstorm event. A study from several Swiss headwater catchments is pre-sented which identified typical patterns in the relations of the magnitude and frequency of torrential sediment events.

Preface

This book is based on lecture notes which were developed for the courses "Natural Hazards" at the University of Natural Resources and Life Sciences (BOKU), Vienna, Austria, and "Management of Torrents and Hillslopes" at the Swiss Federal Institute of Technology, Zurich, Switzerland. The quantitative assessment of channel processes in torrents has always been an important topic of my professional activities, both in research and in teaching, while I was employed at the University of Natural Resources and Life Sciences in Vienna and at the Swiss Federal Institute for Forest, Snow and Landscape Research (WSL) in Birmensdorf, Switzerland. Without the support of these institutions, this book would not have been realized. Some parts of earlier versions of the German course notes were developed in collaboration with my BOKU colleagues Dr. Michael BRAUNER and Dr. Roland KAITNA, and the most recent German versions were checked by my WSL colleagues Christian RICKLI and Christoph GRAF. This book is essentially a translation of the following document published in German, with some updates and minor modifications: "Methoden zur quantitativen Beurteilung von Gerinneprozessen in Wildbächen", Swiss Federal Institute for Forest, Snow and Landscape Research, WSL Berichte, Nr. 9, 105p. (http: www.wsl.ch/publikationen/pdf/13549.pdf). I also would like to acknowledge the translation work of Edward G. PRATER, Bubikon, Switzerland. Peter DAVIES, University of Dundee, UK, supported the publication in his role as IAHR Monograph Series Editor and made a last check of the English text.

Chapter 1

Introduction

1.1 TORRENT PROCESSES AND HAZARD ASSESSMENT

In European countries the term "torrent" refers typically to steep channels in Alpine headwater catchments, with channels steep enough that debris flows can occur apart from fluvial sediment transport. According to this definition, such catchments are associated with drainage areas of less than about 25 km² and, typically, with mean channel gradients steeper than 10% (RICKENMANN & KOSCHNI 2010). This term has not often been used in the recent English scientific literature. As pointed out by SLAYMAKER (1988), the English expression "debris torrent" (originally used in the US Pacific Northwest) was associated with a debris-flow event, which is in contrast to the European meaning of the term referring to a steep stream channel.

Torrent processes in steep channels have their rightful place among the various alpine natural hazards and the corresponding control measures have a long tradition in the European alpine countries. In the planning and execution of such measures, professional experience has been of paramount importance. This experience was based primarily on observations made in earlier torrent events as well as on regular field visits in the catchments of a steep mountain stream. Quantitative measurements, e.g. of the discharge and of the eroded and deposited solid materials, have been increasingly carried out from about the 1990s. Thus, from before that time, there are very limited data on quantitative methods to describe channel processes.

In the meantime, the assessment of torrent processes is based increasingly and primarily on quantitative approaches but also on numerical simulation models. The quantitative description of the channel processes in torrents is based in many cases on earlier and comprehensive investigations to describe similar processes in relatively flat channels or larger catchments. Thus, for example, for about one hundred years systematic investigations on bedload transport have been carried out based on measurements in hydraulics testing laboratories and in natural channels. In more recent times, systematic measurements of bedload transport in steep channels have and are being carried out both in the laboratory and in the field. These studies help improve our knowledge of the processes involved in torrents and show to what extent earlier methods can be adopted or adapted. In this context it was shown, for example, that the flow behavior or the hydraulics in steep channels exhibit differences compared with the behavior in flatter channels, which have to be taken into account in the analysis of bedload transport in steep channels. Likewise, in studying the behavior of debris flows, systematic quantitative measurements using automatic monitoring devices have only been performed in Europe since the 1990s.

In Switzerland, new legal requirements came into force in the 1990s to deal with natural hazards (BUWAL/BWW/BRP, 1997; BWW/BRP/BUWAL, 1997). To produce, for example, hazard maps and flood protection concepts, the process intensities and potential hazard areas have to be quantified. An examination of technical reports on hazard maps and flood protection concepts in the case of torrents and bedload transport processes showed that a comparative assessment is often difficult since, for the production of hazard maps (and, especially, for the process assessment), very different methods are sometimes employed. Furthermore, the methods applied and the basic data, such as the reported input or model parameters used, are, unfortunately, sometimes insufficiently documented. After KIENHOLZ (1999) and KIENHOLZ et al. (2002) technical correctness and clarity in reporting (i) the selected methods and (ii) the assumptions made are the two most important requirements of technical reports with regard to the assessment of natural hazards. The choice and documentation of the procedures adopted is admittedly difficult, since there are few comprehensive surveys of existing methods for the process assessment of natural hazards in torrents–above all, regarding the processes of bedload transport and debris flow. The aim of the present publication is to support the quantitative description of torrent processes as well as the determination of the key parameters and the choice and documentation of the methods.

1.2 ON THE CONTENTS OF THE PRESENT PUBLICATION

Chapter 2 considers the flow resistance in gravel-bedded streams and torrents. It is shown, thereby, that, in the case of steep channels with small flow depths compared with a characteristic size of the surface bed sediment (including clasts up to boulder size), the flow resistance increases considerably and the well-known approaches for flat channels (MANNING-STRICKLER equation, logarithmic flow law) are no longer applicable. The chapter also report show the additional flow resistance can be taken into account quantitatively with regard to bedload transport calculations.

In chapter 3, which considers fluvial bedload transport, the behavior of torrents is introduced and the differences compared with mountain streams are pointed out. Methods are then presented for the characterization of the granular material in the stream bed, such as its grain size distribution. In the analysis of the bedload transport formulae the quantification of the following three main elements is discussed: start of transport, bedload transport rate and consideration of energy losses due to the additional flow resistance in steep channels. In addition, the problem of driftwood debris is discussed, as well as possible hazard locations as a result of erosion or deposition.

Chapter 4 considers firstly the properties of debris flows and presents important elements of process and hazard assessment. The conditions under which debris flows can occur are then discussed. Quantitative empirical or semi-empirical methods are presented, with the determination of important parameters of the flow and depositional behavior. In the presentation of simulation models, the focus is on the flow and depositional behavior on the torrent's fan. Finally, the importance of scenarios of torrent processes and of traces of deposition on the fan is indicated.

In chapter 5 the magnitude and frequency of torrent events are considered. The importance of a field-based estimate of the size of the event is emphasized. Following this, a combined method for estimating the size of an event is presented. Finally, in the discussion of the frequency of torrent events, attention is drawn to the considerable importance of historical data, with a presentation of an important study of debris flow activity in typical Swiss torrent catchments.

Chapter 6 contains a short summary of important aspects that should be taken into account in the hazard assessment of torrent processes.

Chapter 2

Flow resistance in gravel-bedded streams and torrents

Flow resistance is a measure of the friction between the water and the base of the channel and the slopes of the banks. For a given flow depth or discharge, flow resistance laws allow the determination of the mean flow velocity in the channel as a function of the channel geometry and the bed roughness. The quantification of the flow resistance is also important for the calculation of bedload transport.

2.1 LOGARITHMIC FLOW LAW

For the determination of the flow behavior in open channels, the universal logarithmic flow law, developed by COLEBROOK-WHITE for the hydraulics of pipe flow, and the DARCY-WEISBACH friction coefficient can be applied (CHANSON, 2004). In the universal flow resistance law Eq. 2.1 with the VON KARMAN constant, the logarithmic part contains a quotient of an integration constant times the flow depth and the equivalent roughness k_s. As Fig. 2.1 shows, the vertical velocity distribution in the case of steep and rough channels (left) sometimes deviates substantially from the logarithmic velocity law (right). Different modifications to the original distribution law attempt to take this phenomenon into account.

$$\sqrt{\frac{8}{f}} = \frac{1}{\kappa}\ln\left(\frac{a\,h}{k_s}\right) \tag{2.1}$$

$$\frac{V}{v^*} = \frac{V}{\sqrt{ghS}} = \sqrt{\frac{8}{f}} \tag{2.2}$$

Here f = friction coefficient after DARCY-WEISBACH as defined by Eq. 2.2, κ = VON KARMAN coefficient (= 0.4), h = flow depth, k_s = equivalent roughness height ("sand roughness"), a = coefficient (frequently a = 12), V = mean flow velocity, g = gravitational acceleration, S = channel slope (or friction slope) (in all equations in this publication S is expressed in the unit [m/m] and not in [%]), $v^* = (ghS)^{0.5}$ = shear velocity, k_s = const. D_x, and D_x = characteristic grain size for which $x\%$ of the material is finer. The characteristic grain size of the channel bed refers in all flow formulae to the grain size distribution of the surface material or of the armor layer. In gravel-bed streams the relative flow depth is typically defined as h/D_{84}.

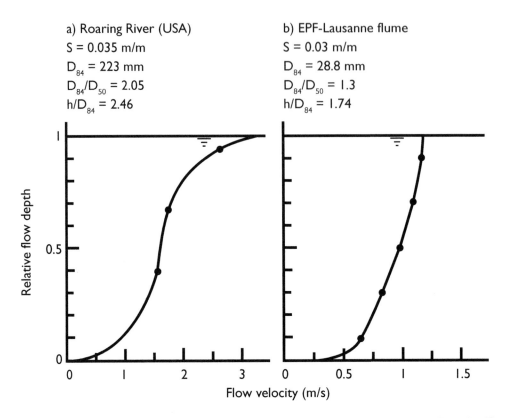

a) Roaring River (USA)
S = 0.035 m/m
D_{84} = 223 mm
D_{84}/D_{50} = 2.05
h/D_{84} = 2.46

b) EPF-Lausanne flume
S = 0.03 m/m
D_{84} = 28.8 mm
D_{84}/D_{50} = 1.3
h/D_{84} = 1.74

Figure 2.1 Velocity distribution in steep channels for shallow flows (small relative flow depth). (a) Non-uniform grain size distribution, Roaring River, Colorado, USA. (b) Uniform grain size distribution, laboratory flume at EPF Lausanne. Modified from BATHURST (1993).

Table 2.1 Empirical derivation of the equivalent sand roughness k_s for sand/gravel beds, gravel beds and for rough channels.

$k_s = D_{65}$	EINSTEIN (1942)	sand/gravel bed
$k_s = D_{90}$	GARBRECHT (1961)	sand/gravel bed
$k_s = 3.5\ D_{84}$	HEY (1979)	gravel bed
$k_s = 4.5\ D_{50}$	THOMPSON & CAMPELL (1979)	cobble/boulder bed

As equivalent roughness k_s, the natural surface roughness is understood as being given by "grains of constant size with the densest possible packing" (SCHRÖDER 1994). It should be noted that, according to this definition, the actual grain size is only of secondary importance for the determination of the grain roughness. Some empirically-determined relations between k_s and D_x are listed in Table 2.1.

The flowing water separates at the roughness elements of the bed and the turbulence that is thereby initiated finally disappears due to energy dissipation. The bed structure is also of importance for the determination of the flow resistance, besides the height of the roughness elements k_s. Thus, a derivation of k_s from the characteristic grain size diameter can only be determined from empirical, functional relationships and subsumed values of geometrical roughness, packing and arrangement of the roughness elements. These functional relationships can only, therefore, give an indication of the actual physical behavior.

For the determination of the flow resistance, based on the logarithmic resistance law, various formulations have been proposed for gravel-bed streams (KEULEGAN 1938; HEY 1979; BATHURST 1985; SMART & JÄGGI 1983).

$$\frac{V}{v^*} = 6.25 + 5.62 \log\left(\frac{h}{k_s}\right)$$

$$(2.3)$$
KEULEGAN (1938)

$$\frac{V}{v^*} = 4 + 5.62 \log\left(\frac{h}{D_{84}}\right)$$

$$(2.4)$$
BATHURST (1985), natural gravel,
plane bed, $0.4\% < S < 9\%$, $h/D_{84} <$ approx. 7–10

$$\frac{V}{v^*} = 5.75\sqrt{1 - e^{\left[-0.05\frac{h}{D_{90}\sqrt{S}}\right]}} \log\left(8.2\frac{h}{D_{90}}\right)$$

$$(2.5)$$
SMART & JÄGGI (1983),
flume, plane bed, $S < 20\%$

As many investigations have shown, there is a functional relationship between flow depth, equivalent roughness (k_s) and channel slope. In particular, the resistance behavior changes at a relative flow depth of about 4 to 5. After DITTRICH (1998), a layer with a reduced velocity distribution develops in the channel-water contact region above very rough beds. The layer is called the roughness sub-layer. This situation results in an S-shaped deformation of the vertical velocity distribution, which results in an increased stabilizing velocity reduction in the region of the bed and thus a reduction of shear stress, but at a certain distance from the bed the velocity increases (ROSPORT 1998). BEZZOLA (2002) attempted to improve the description of the resistance behavior of the bed through an increased influence of the height y_R. of the lower roughness layer After BEZZOLA (2002), the lower roughness layer depends on the shape, density and exposure of the roughness elements, but not on the relative flow depth; the author proposed to quantify its thickness as a function of D_{90} (Table 2.2).

In general, in the case of narrow channel cross-sections with $W/h <$ approx. 10 (W = channel width), the flow depth is often replaced by the hydraulic radius R, such that the wall or bank friction is taken into account. The hydraulic radius R is

Table 2.2 Thickness of the lower roughness layer after BEZZOLA (2002).

Roughness element	Height of the lower roughness layer y_R
Uniform size, spherical	$0.5\,D_{90}$
Uniform size, natural particles	$1\,D_{90}$
Variabel size, natural particles	$2\,D_{90}$

defined as $R = A/P$, with A = flow cross-section and P = wetted perimeter of the flow cross-section. In the flow laws presented in chapter 2, the hydraulic radius R is then used instead of the flow depth, and in this case, the shear velocity is calculated as $v^* = (gRS)^{0.5}$.

2.2 EMPIRICAL FLOW RESISTANCE LAWS (POWER LAWS)

The most widely-known empirical flow formula is that of GAUCKLER-MANNING-STRICKLER (HAGER 2001), based on studies of GAUCKLER (1867), MANNING (1890), and STRICKLER (1923). The associated flow resistance laws are in the form of power laws, thus being similar to the definition of the DARCY-WEISBACH friction coefficient (Eq. 2.2). The STRICKLER coefficient k_{St} is dimensional and is usually used for the characterization of the total resistance, which takes into account grain and form roughness as well as any additional roughness. In the case of a large influence of the grain roughness (i.e. a straight channel with a plane bed surface), there is a strong correlation with the equivalent sand roughness k_s.

$$V = k_{St} R^{2/3} S^{1/2}$$

(2.6)
STRICKLER (1923)

$$k_{st} = \frac{21.1}{\sqrt[6]{\varepsilon_0}}$$

(2.7)
STRICKLER (1923)

$$k_{st} = \frac{26}{\sqrt[6]{D_{90}}}$$

(2.8)
MEYER-PETER & MÜLLER (1949)

$$k_{st} = 6.7\sqrt{g}\,\frac{1}{\sqrt[6]{k_s}} \approx \frac{21.1}{\sqrt[6]{\varepsilon_0}}$$

(2.9)
after STRICKLER (1923)

$$c = \frac{V}{v^*} = \sqrt{\frac{8}{f}} = 6.7 \cdot \left(\frac{R}{D_{90}}\right)^{1/6}$$

(2.10)
STRICKLER (1923) with Eq. 2.9 and $k_s = D_{90}$

The STRICKLER formula was obtained from tests with relative flow depths $h/D > 10$. If $\varepsilon_0 = D_{90}$ is inserted into Eq. 2.9 as the roughness height, then Eq. 2.6 can be transformed into Eq. 2.10. A comparison of Eq. 2.10 with the logarithmic flow law in Fig. 2.2 shows that, for shallow flow depths, the STRICKLER formula clearly exhibits a different trend from that of the logarithmic flow law. For Eq. 2.6, Eq. 2.7, Eq. 2.8, Eq. 2.9, the value R is in [m], ε_0, D_{90} and k_s are in [m], k_{St} is in [m$^{1/3}$/s] so that V has the unit [m/s]. Eq. 2.10 describes a medium flow resistance in alpine gravel-bed streams with relative flow depths $R/D_{90} >$ approx. 10 (cf. also Eq. 2.21 in chapter 2.4).

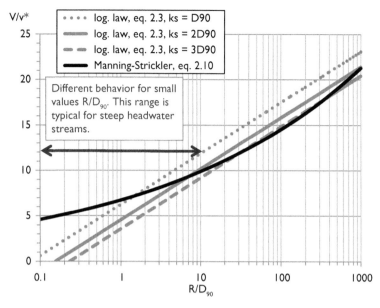

Figure 2.2 Comparison of different flow resistance formulae, presented as normalized flow velocity V/v^* as a function of the relative flow depth, R/D_{90}. The empirical STRICKLER formula Eq. 2.10 exhibits a very different behavior for small relative flow depths from that for the logarithmic flow laws such as, e.g., KEULEGAN (1938) (Eq. 2.3). Modified from BEZZOLA (2005).

Table 2.3 STRICKLER coefficients k_{St} for the total roughness of natural channels after ZELLER & TRÜMPLER (1984).

	$k_{St}\ [m^{1/3}/s]$
Torrents (steep headwater streams)	
Coarse gravel bed with cobbles, straight	20–25
Coarse gravel bed with cobbles, winding	15–20
Stone bed with individual boulders, straight	12–17
Boulder bed, step-pool or rapid-pool, irregular	8–15
Boulder bed, step-pool or rapid-pool, irregular with vegetation	5–12
Mountain rivers	
Gravel and cobble bed, straight	20–33
Cobble bed with boulders, straight	14–25
Boulder bed, straight	10–15

For torrents, with typically small relative flow depths, little is known about suitable values of the equivalent sand roughness k_s in the logarithmic flow law or any reasonable modifications of this flow law. The STRICKLER formula was, therefore, often used in the past with typical values of the STRICKLER coefficients k_{St} as given in Table 2.3.

Many field studies (JARRET 1984; HODEL 1993; BATHURST 1985; RUF 1990; RICKENMANN 1994, 1996; ZELLER 1996; PALT 2001) and also laboratory flume studies

(MEYER-PETER & MÜLLER 1948; ROSPORT 1998) showed that a marked change in the resistance behavior is observed for a channel slope of more than approximately 1–3%. This phenomenon is due to the effect of a greater presence of distinct morphological bed structures (macro-roughness) at steeper slopes, as well as to the influence of smaller relative flow depths. In natural channels, for channel slopes of more than approximately 1–3%, discharges frequently occur with relative flow depths h/D or R/D smaller than 3–5.

RICKENMANN (1994, 1996) developed formulae for flow velocity as a function of discharge, channel slope and a characteristic grain size, based on 373 field measurements (see Eq. 2.11, Eq. 2.12, with Q = channel discharge). The formulae are dimensionally correct and apply to natural channel sections with bed slopes between 0.01 and 63%. The partitioning into two domains with the dividing point $S = 0.8\%$ reflects the situation mentioned above, that, above approximately 1% channel slope, there is an increased flow resistance due to pronounced bed structures. The equations are analogous to approaches following hydraulic geometry theory (GRIFFITHS 2003; SINGH 2003, PARKER et al., 2007; EATON 2013). Flow velocities calculated according to Eq. 2.11 and Eq. 2.12 are compared further below with independent velocity measurements (Fig. 2.6), along with other more recent approaches discussed in the following subchapter.

$$V = \frac{0.37 g^{0.33} Q^{0.34} S^{0.20}}{D_{90}^{0.35}}$$

(2.11)

RICKENMANN (1996), $0.8\% \leq S < 63\%$

$$V = \frac{0.96 g^{0.36} Q^{0.29} S^{0.35}}{D_{90}^{0.23}}$$

(2.12)

RICKENMANN (1996), $0.01\% < S < 0.8\%$

2.3 VARIABLE POWER LAW

In the domain of relative flow depths of h/D_{84} and R/D_{84} (or R/D_{90}) below a value of approximately 10, the STRICKLER formula exhibits a rather different form from the logarithmic flow law (Fig. 2.2, Fig. 2.3). On the other hand, simple logarithmic flow laws result partially in too small or even negative flow velocities for h/D_{84} < approx. 1.

RICKENMANN & RECKING (2011) compared six flow formulae with a total of 2890 worldwide field measurements of flow velocities in gravel-bed streams. This data set also includes many measurements for steep streams. The best description of the average trend of all data was achieved using the variable power equation (VPE) of FERGUSON (2007):

$$\frac{V}{v^*} = \sqrt{\frac{8}{f}} = \frac{a_1 a_2 (h/D_{84})}{\sqrt{a_1^2 + a_2^2 (h/D_{84})^{5/3}}}$$

(2.13)

FERGUSON (2007)

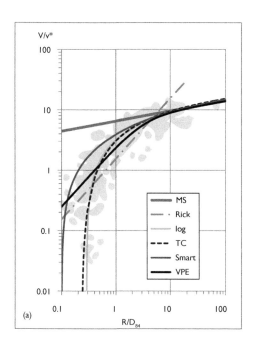
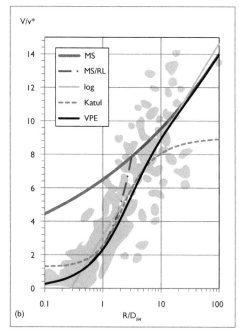

Figure 2.3 (a) Double logarithmic plot and (b) Semi-logarithmic plot of $(8/f)^{0.5}$ $(= V/v^*)$ versus relative flow depth R/D_{84}, with 376 field measurements. The colored lines represent different flow laws: MS = Manning-Strickler Eq. (valid for large h/D_{84} values), RL = roughness layer Eq. (valid for small h/D_{84} values), MS/RL = "best" combination, log = logarithmic Eq. HEY (1979), Katul = KATUL et al. (2002), Rick = RICKENMANN (1991), Smart = SMART et al. (1992), TC = THOMPSON & CAMPELL (1979), VPE = variable power equation of FERGUSON (2007). Modified from FERGUSON (2007).

Here, the coefficient values $a_1 = 6.5$ and $a_2 = 2.5$ were used. With the aid of two dimensionless parameters for the flow velocity, U^{**}, and for the unit discharge, q^{**}, Eq. 2.13 with $a_1 = 6.5$ and $a_2 = 2.5$ can be presented alternatively as follows:

$$U^{**} = 1.443\, q^{**0.60} \left[1+\left(\frac{q^{**}}{43.78}\right)^{0.8214}\right]^{-0.2435} \tag{2.14}$$

RICKENMANN & RECKING (2011)

$$U^{**} = \frac{V}{\sqrt{gSD_{84}}} \tag{2.15}$$

$$q^{**} = \frac{q}{\sqrt{gSD_{84}^3}} \tag{2.16}$$

In the range of relative flow depths h/D_{84} smaller than approximately 10, the majority of the measurements from torrent-like channel reaches (mostly data of

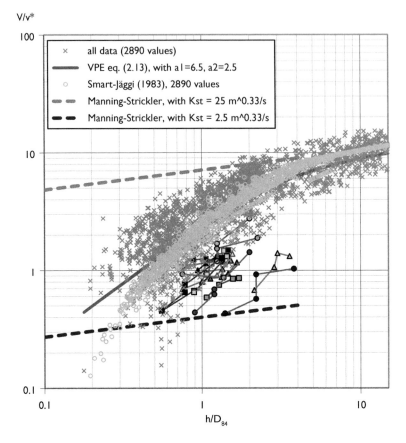

Figure 2.4 Measurements of the flow velocity in natural channels, plotted as V/v^* vs. h/D_{84}, for relative flow depths smaller than 20. VPE = variable power equation, Eq. 2.13. The data points connected with solid lines refer to observations of DAVID et al. (2010) for step-pool and cascade reaches with channel slopes $S = 0.06$ to 0.18; they are from torrent-like channel reaches and generally lie below the VPE line. Based on data from RICKENMANN & RECKING (2011).

DAVID et al. (2010) in Fig. 2.4) exhibit much larger DARCY-WEISBACH friction coefficients f (or a smaller coefficient a_2) than the average trend of the remaining data according to Eq. 2.13 (Fig. 2.4) or Eq. 2.14 (Fig. 2.5). The change in the flow velocity (V/v^*) with relative flow depth (h/D_{84}) for these data (shown by the colored connecting lines per channel reach) fits better with Eq. 2.13 than with the STRICKLER formula. For relative flow depths h/D_{84} smaller than approximately 4, Eq. 2.13 can be approximated by:

$$\frac{V}{v^*} = \sqrt{\frac{8}{f}} = 2.2 \frac{h}{D_{84}} \tag{2.17}$$

According to the data of DAVID et al. (2010) the coefficient a_2 in very rough channels can be reduced to about 0.4; i.e. the flow velocity compared with the average

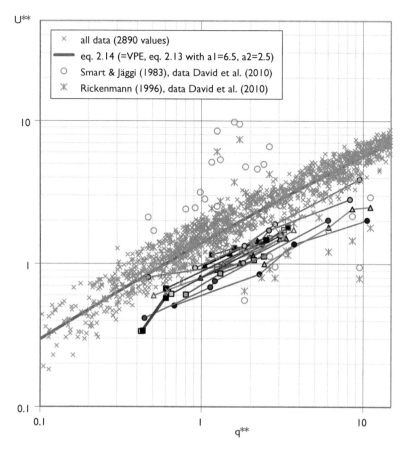

Figure 2.5 Measurements of the flow velocity in natural channels, plotted as U^{**} vs. q^{**}, for smallrelative flow depths ($q^{**} < 20$). VPE = variable power equation in the form of Eq. 2.14. The data points connected with solid lines refer to observations of DAVID et al. (2010) for step-pool and cascade reaches with channel slopes $S = 0.06$ to 0.18; they are from torrent-like channel reaches and generally lie below the VPE line. The flow laws of SMART & JÄGGI (1983) (Eq. 2.5) as well as of RICKENMANN (1996) (Eq. 2.11) tend to over- and partly under-estimate the observations of DAVID et al. (2010). Based on data data from RICKENMANN & RECKING (2011).

trend of the other data can be reduced approximately by a factor 5 to 6. A reason for this could be that, in the case of the channels investigated by DAVID et al. (2010), large wood pieces were caught in the stream's bed structures, thereby increasing the roughness. When the flow laws of SMART & JÄGGI (1983) (Eq. 2.5), as well as of RICKENMANN (1996) (Eq. 2.11), are applied to the data from the torrent-like channel reaches of David et al. (2010) presented in Fig. 2.5, they show a tendency to overestimate the flow velocity in such conditions. A further comparison of the flow formulae discussed here was made using independent measurements of the flow velocity in mountain rivers in the Himalayas (PALT 2001); it is likewise shown that generally the best agreement is obtained with the VPE solutions Eq. 2.13 and Eq. 2.14 (Fig. 2.6).

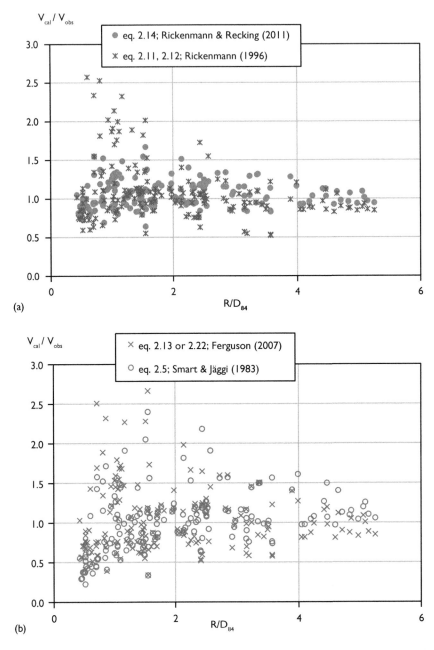

Figure 2.6 Comparison of the calculated flow velocity (V_{ber}) with different flow laws and with independent measurements of the flow velocity (V_{gem}) in mountain rivers in the Himalayas (PALT 2001): (a) RICKENMANN (1996), Eq. 2.11 and Eq. 2.12; RICKENMANN & RECKING (2011), Eq. 2.14; (b) FERGUSON (2007), Eq. 2.13 or Eq. 2.22, calculated with *R*; SMART & JÄGGI (1983), Eq. 2.5, calculated with *R*.

The VPE flow resistance formula Eq. 2.13 with the coefficients $a_1 = 6.5$ and $a_2 = 2.5$ was also applied to some steep torrents (channel sections without wood debris) in Switzerland and resulted in quite good agreement with field measurements for mean flow velocity (NITSCHE et al., 2012a).

In summary, more recent investigations revealed that in steep and rough channels (with $h/D_{84} < 4$) the flow resistance can be better described with the power law based on FERGUSON (2007) than with the STRICKLER formula or with a logarithmic flow law. However, the roughness coefficient a_2 in this power law is also strongly dependent on the channel morphology. It is well known that the STRICKLER value for torrents can vary within a range of about 2 $m^{1/3}/s$ up to about 30 $m^{1/3}/s$ (c.f. Table 2.3); however, the MANNING-STRICKLER formula does not correctly predict the increase of $V/v*$ with increasing h/D_{84} (Fig. 2.4). It should also be taken into account that (i) the flow behavior in torrents may be difficult to approximate by a one-dimensional approach, (ii) frequently there may be local changes between sub- and supercritical flow and (iii) the flow conditions generally depend strongly on the discharge or the relative flow depth. Many of these aspects still require further detailed study.

2.4   PARTITIONING OF THE FLOW RESISTANCE

Total flow resistance of surface runoff in a channel consists of the skin friction of the water flowing along the individual grains of the bed (grain roughness) and the friction losses resulting from bed forms, large immobile grains and irregular channel geometry (macro-roughness). In steep channels, the additional friction losses (apart from grain roughness) may be due to form drag, local flow accelerations and hydraulic jumps and the formation of multiple flow paths in shallow flows characterized by low relative flow depths h/D. Earlier concepts of flow resistance partitioning distinguished between grain and form roughness. A correction term to account for roughness losses in the calculation of bedload transport was introduced, for example, in the approaches of MEYER-PETER & MÜLLER (1948) and PALT (2001), where the relation between grain roughness (k_r) and total roughness (k_{St}) is important. A similar correction is introduced here by firstly using the Manning coefficient (n), which is the reciprocal of the Strickler coefficient $(n = 1/k_{St})$.

Based on 373 measurements of mean flow velocity in steep channels (RICKENMANN, 1994, 1996), a correction factor was determined to partition the flow resistance into the base-level roughness (n_o) and the total roughness (n_{tot}). This correction factor is expressed as a function of either total discharge Q in Eq. 2.18 after RICKENMANN et al. (2006a) or of flow depth h in Eq. 2.19 after Chiari et al. (2010) (see also Fig. 2.7). These equations for the partitioning of flow resistance are implemented in the sediment transport simulation program SETRAC (CHIARI & RICKENMANN 2011) and in the first version of the follow-up program TOMSED (www.bedload.at).

$$\left(\frac{n_o}{n_{tot}}\right) = \frac{0.131\, Q^{0.19}}{g^{0.096}\, S^{0.19} D_{90}^{0.47}}$$

(2.18)
RICKENMANN et al. (2006a)

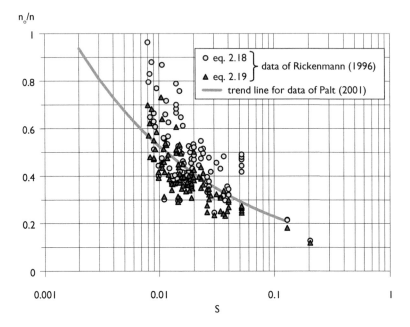

Figure 2.7 The correction factor (n_o/n_{tot}) as a function of the channel slope S, to take into account energy losses due to high flow resistance in steep channels. Modified from CHIARI et al. (2010).

$$\left(\frac{n_o}{n_{tot}}\right) = 0.092 \left(\frac{b}{D_{90}}\right)^{0.33} S^{-0.35} \tag{2.19}$$

CHIARI et al. (2010)

The partitioning of the flow resistance can also be carried out using the DARCY-WEISBACH coefficient f by dividing the total friction (f_{tot}) into a base-level friction (f_o) and an additional friction component (f_{add}):

$$f_{tot} = f_0 + f_{add} \tag{2.20}$$

Most bedload transport equations are based on laboratory flume experiments, in which the form or macro-roughness losses were negligible and the flow conditions were associated with relative flow depths mostly larger than about 7–10. Thus the MANNING-STRICKLER equation such as Eq. 2.10 or Eq. 2.21 provides a good quantification of the mean flow resistance in deeper flows, which is considered here as the base-level resistance (f_o). However, additional flow resistance (f_{add}) is present in steep streams with coarse roughness elements such as immobile boulders and step-pool sequences and small relative flow depths less than about 10 (e.g. torrent channels).

RICKENMANN & RECKING (2011) used the VPE approach of FERGUSON (2007) (Eq. 2.13) to develop a method to partition the flow resistance for conditions with medium to large-scale roughness (in the sense of BATHURST et al., 1981; here for

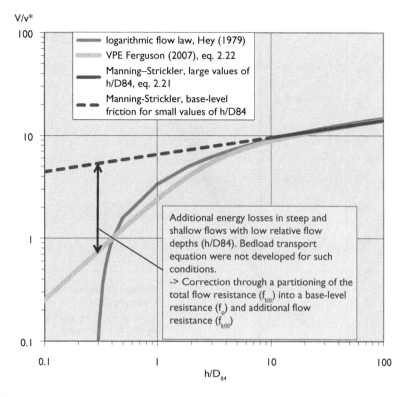

Figure 2.8 Partitioning of the flow resistance for medium and large-scale roughness conditions (here h/D_{84} < approx. 10). Based on a formula of the type MANNING-STRICKLER, a base level for the flow resistance is determined (dashed purple line), corresponding to the flow conditions in the case of small-scale (here h/D_{84} > approx. 10) roughness (solid purple line), after the approach of RICKENMANN & RECKING (2011).

h/D_{84} < approx. 7). Based on a MANNING-STRICKLER type formula, a base-level flow resistance is calculated (Fig. 2.8), which corresponds to flow conditions with small-scale roughness (here for h/D_{84} > approx. 7):

$$\frac{V_o}{v^*} = \sqrt{\frac{8}{f_o}} = 6.5 \left(\frac{h}{D_{84}} \right)^{1/6} \tag{2.21}$$

If Eq. 2.21 is used for medium- to large-scale roughness conditions, an appropriate level for the base-level friction can be calculated for f_o or for the virtual flow velocity V_o. The total flow resistance is determined with the VPE as follows:

$$\frac{V_{tot}}{v^*} = \sqrt{\frac{8}{f_{tot}}} = \frac{(6.5)\,(2.5)\,(h/D_{84})}{\sqrt{6.5^2 + 2.5^2 (h/D_{84})^{5/3}}} \tag{2.22}$$

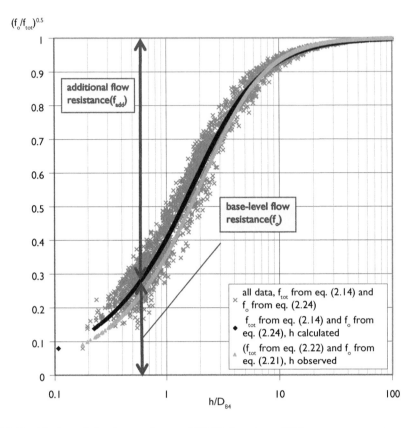

Figure 2.9 Partitioning of the flow resistance $(f_o/f_{tot})^{0.5}$, based on 2890 measurements in gravel-bed streams and partly in torrents. The values $(f_o/f_{tot})^{0.5}$ are essentially a function of the relative flow depth. The dark red line corresponds to calculation according to Eq. 2.25, while the light purple line corresponds to calculation according to Eq. 2.23. Modified after RICKENMANN & RECKING (2011).

Thus, the ratio of base-level to total flow resistance can be calculated as:

$$\sqrt{\frac{f_o}{f_{tot}}} = \frac{V_{tot}(h)}{V_o(h)} \qquad (2.23)$$

The above proposed partitioning of the flow resistance is basically a function of the relative flow depth (Fig. 2.9). It is an all-inclusive and empirical approach, but it implicitly contains information about an average increase of roughness in steep channels with irregular bed morphology and shallow flows. Instead of calculating (f_o/f_{tot}) as a function of flow depth with Eq. 2.21 to Eq. 2.23, (f_o/f_{tot}) can be determined as a function of the unit discharge q as follows:

$$U_o** = \frac{V_o}{\sqrt{gSD_{84}}} = 3.704 \left(q**\right)^{0.4} \tag{2.24}$$

Together with Eq. 2.16, $V_o(q)$ is then obtained. For the calculation of the total resistance f_{tot} or V_{tot} Eq. 2.14 together with Eq. 2.15 and Eq. 2.16 is used, which then gives $V_{tot}(q)$. The ratio of base-level to total flow resistance is then given as:

$$\sqrt{\frac{f_o}{f_{tot}}} = \left(\frac{V_{tot}(q)}{V_o(q)}\right)^{1.5} \tag{2.25}$$

The ratio (f_o/f_{tot}) represents a reduction factor, which is a measure of that part of the total flow energy (or total bed shear stress) available for bedload transport. The partitioning of the flow resistance by means of the value $(f_o/f_{tot})^{0.5}$ is the basis for determining a reduced energy slope S_{red}, which is then introduced into the calculation of the bedload transport (see chapter 3.4.4).

Chapter 3

Fluvial bedload transport

3.1 CHARACTERIZATION OF TORRENTS AND MOUNTAIN RIVERS

In mountain rivers and torrents, hydrologic and hydraulic processes are characterized by extreme variability, both in the spatial and the temporal domains. The main causes of this lie in the strong interaction of geological constraints, earth surface processes, and the channel network in alpine catchments (HASSAN et al., 2005a; COMITI & MAO 2012; CHURCH 2013). This results in a high variability of the following parameters:

- sediment supply or availability and sediment transport
- composition of the grain size distribution of the stream bed and of the source areas
- channel geometry along the stream and in the lateral direction
- highly variable (but generally low) runoff depths
- flow behavior in the transition regions subcritical-supercritical-subcritical flow

Typical torrent channels are greatly influenced by these parameters so that a consideration of the geological and morphological conditions in and along the channel is of great importance. Sediment is often supplied to the channel by colluvial processes, while the channel bed may consist partly of bedrock and, thus, may have only a semi-alluvial character. Mountain rivers, in contrast, typically have an alluvial streambed that reflects a single dominant formation process. In European countries, torrents refer typically to Alpine catchments with channels steep enough that debris flows can occur in addition to fluvial sediment transport. According to this definition, such catchments are associated with drainage areas of less than about 25 km² (RICKENMANN & KOSCHNI 2010; MARCHI & BROCHOT 2000; MARCHI & D'AGOSTINO 2004). Typical differences between torrents and mountain rivers are summarized in Fig. 3.1.

An important difference between torrents and alluvial mountain rivers concerns both sediment supply or sediment availability and the runoff conditions. Typically, torrents have sediment supply-limited conditions, whereas bedload movement in mountain rivers is mostly limited by the hydraulic transport capacity (MONTGOMERY &

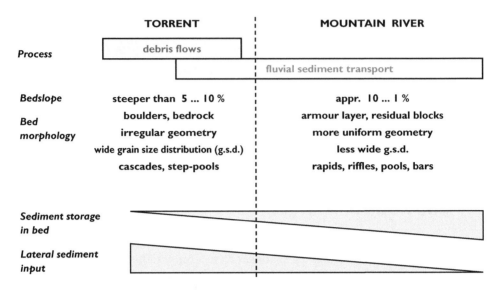

Figure 3.1 Overview of some differences between torrents and alluvial mountain rivers. The gray areas signify, from left to right, the increasing and decreasing importance, respectively, of sediment storage in the bed and of lateral sediment input with increasing catchment size.

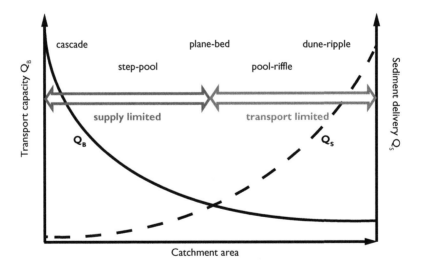

Figure 3.2 Transport-limited and sediment supply-limited conditions in mountain catchments. With increasing drainage area the channel slope decreases. The typical morphological structures also vary with the channel slope. Modified after MONTGOMERY & BUFFINGTON (1997).

BUFFINGTON 1997; Fig. 3.2). In torrent catchments, peak water runoff and also the formation of debris flows is often triggered by convective rainstorms with high precipitation intensities and short durations. These storms are associated with a rapid increase in discharge, and flood events typically lasting less than a few hours. Longer duration rainfall events may be more important in mountain rivers, resulting in a more gradual increase in runoff conditions.

Three separate morphological elements can be differentiated along a stream channel: streambed, bank and the adjoining hillside. The morphology of the bed and the banks of a mountain river is closely associated with the runoff and bedload transport processes. In its lower parts a mountain river in its natural environment can adjust its course or total width by meandering, bar formation and braiding, reflecting the interactions between runoff, bedload transport, grain sizes and the general valley slope (MONTGOMERY & BUFFINGTON 1997). In contrast, if a torrent or mountain river is constrained laterally by the valley slopes or bedrock outcrops, morphological structures in the vertical dimension (e.g. step-pools and cascades) become more prominent (Fig. 3.1, Fig. 3.2). The formation of these structures may be favored by large immobile boulders or large wood fragments, and they are associated with a high energy dissipation of flowing water. A destruction of steps may lead to a temporary increase of bedload transport (TUROWSKI et al., 2009). The typical morphological structures generally change with mean channel slope or with the size of the catchment area (MONTGOMERY & BUFFINGTON 1997), which is also indicated in Fig. 3.1 and Fig. 3.2.

The channel morphology can be described by the three elements: channel geometry, bed form and grain shape (DE JONG & ERGENZINGER 1995; MORVAN et al., 2008). Considering these structures, previous authors have distinguished between grain resistance and form resistance in bedload transport calculations for gravel-bed rivers (MEYER-PETER & MÜLLER 1948; CARSON 1987; GOMEZ & CHURCH 1989). However, distinguishing between these resistances is only really possible in channels with sandy beds; in gravel-bed streams (and especially in torrent channels), such a distinction is questionable (ZIMMERMANN 2010; RICKENMANN & RECKING 2011).

Colluvial and fluvial stream bed types can be distinguished according to the relative importance of hillslope sediment delivery and hydraulically-forced evolution of the bed sediments (MONTGOMERY & BUFFINGTON 1997). In colluvial stream reaches, sediment delivery from talus slopes is dominant, and along–channel sediment transfer occurs mainly due to debris flows or very rare and extreme flood events. Colluvial reaches are, therefore, characterized by angular, unsorted particles. Reaches dominated by fluvial bedload transport are characterized by rounded, well-sorted sediment particles, and often a coarse armor layer is formed with a preferred orientation of the particles depending on flow direction (ABERLE & NIKORA 2006). A torrent channel is often composed of both colluvial and fluvial sediments, and an armor layer is not necessarily formed, as is often the case in mountain rivers with purely fluvial sediments. It is, therefore, questionable to what extent concepts regarding the formation and breakup of an armor layer are also applicable in torrent channels.

Further important characteristics of steeper channel reaches in mountain streams and torrents are longitudinal bed structures that are developed. Typical morphologies for stream beds with slopes steeper than about 3% are step-pool sequences (Fig. 3.3) and cascades (MONTGOMERY & BUFFINGTON 1997; GRANT et al., 1990). Empirical studies show that such structures may be stable for a flood event with return periods on the order of about 50 years (CHIN 1989; LENZI et al., 2004).

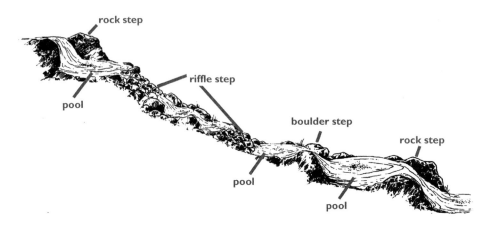

Figure 3.3 Step-pool sequences as typical bed structures in steeper streams and torrent channels. Adapted from HAYWARD (1980).

3.2   SEDIMENTOLOGICAL PARAMETERS

The grain size distribution and the grain shape have a large influence on the transport of sediments in alpine watercourses. Due to the large spatial and temporal variability of the sediment distribution, it is a challenging task to determine a representative grain size distribution for a given channel reach.

For the computation of channel discharge and sediment transport, characteristic grain diameters are required to determine the flow resistance (e.g. D_{84}, D_{90}), the start of mobilization (e.g. D_{50}, D_{65}, and possibly grain shape) and the transport efficiency as a function of grain size distribution parameters (D_{16}, D_{30}, D_{84}, D_{90}). In terms of a cross-section, the directly-measured grain size distribution represents only a snapshot, the result of the immediate hydraulic conditions (i.e. those that have just taken place) and of the current bedload transport. Such a static snapshot, therefore, only gives a partial picture of the dynamically-changing sedimentological conditions during an event or over a specific period of time.

The spatial variability results from the often very heterogeneous transport and deposition conditions as well as the various sediment sources. Thus, the sedimentology of sediment sources and the channel geometry (separate for bed and bank) should be distinguished. To determine the grain size distribution of the surface layer (armor layer) over a reach, different morphological elements such as steps, pools, rapids and gravel banks should be taken into account, ideally proportionately to their spatial occurrence. However, it is often not easy to clearly separate zones of different stream-bed morphology. The temporal variability depends on the process dynamics during the transport event and can only be taken into account more thoroughly by a comparison of the situation after several transport events. A possible approach is to consider exposed earlier deposits or the comparative evaluation of the bed and the adjacent bank.

Among other factors, the grain shape influences the initial mobilization and the transport process itself. Thus, with an increasing plate-like shape of the individual grain a greater alignment similar to a roof-tile structure of the fluvially-formed bed is possible and, thus, a reduction of the effective grain roughness for the same stable

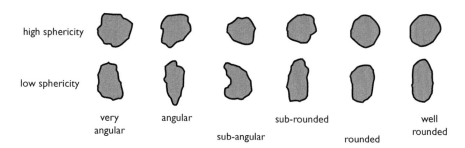

Figure 3.4 Subdivision of the grain shape according to roundness (very angular to very round) and according to sphericity. Modified from: [*http://homepage.usask.ca/~mjr347/prog/geoe118/geoe118.017.html*]

bedding structure. A general classification according to shape and roundness can be carried out following, for example, SCHREINER (1997). Fig. 3.4 gives a description of grain shape according to roundness, and it is also possible to distinguish between differences in sphericity.

3.3 DETERMINATION OF GRAIN SIZE DISTRIBUTION AND OF CHARACTERISTIC GRAIN SIZES

Various analysis methods may be found in the literature to determine the grain size distribution (BUNTE & ABT 2001). These methods have been developed either for a specific grain size spectrum or for specific deposition conditions. In the case of steep channels, typically with a broad grain size spectrum, different methods have to be combined for a suitable analysis. In using such statistical methods it has to be recognized that they were usually derived for different sedimentological conditions and that they have not yet been verified systematically for use in torrent channels. For the purpose of a better comparison, all methods should be based on a unified grain size classification; an overview of common classification schemes is provided in Table 3.1.

3.3.1 Volume analysis of sediment

A certain volume of sediment is collected from the stream bed. In this way the grain size can be determined without the influence of alignment and layering, but the site has to be easily accessible and the sample must consist of granular material. Samples can be taken not only near the surface but also at different depths. The bigger the maximum grain size, the bigger is the volume that has to be tested (CHURCH et al., 1987).

The volumetric analysis can be carried out by means of sieving and weighing the material within each individual grain size class so that different characteristic grain sizes such as the median axis (b-axis) are determined. Sieving can be either wet or dry, but weighing should always be carried out on dry material, since the influence of the weight of water in the case of fine-grained material can be substantial. A few approaches to estimate the necessary sample size are listed in Table 3.2.

Table 3.1 Classification schemes to enable determination of grain size distributions. The typical range of application of the various methods of analysis is depicted by gray bars. ISO: International Organization for Standardization; VSS: Swiss Association of Road and Traffic Engineers; USCS: Unified Soil Classification System; ÖNORM: Austrian Standards Institute. LNA: line by number analysis (see below in this chapter).

ISO Scale		Metric classes	VSS / USCS	ÖNORM B4412	LNA (VAW-ETHZ)	Sieve analysis	Surface analysis	Photo sieving
	(German)	[mm]	[mm]	[mm]	[mm]			
					>2000			
					1500			
large boulder	grosse Blöcke				1200			
					1000			
					800			
		> 630			600			
					500			
					400			
boulder	Blöcke				350			
					300			
					250			
		200 - 630	> 200		200			
					150			
					120			
cobble	Steine				100			
					80			
		63 - 200	60 - 200	63	60			
					40			
coarse gravel	grober Kies			31.5	30			
		20 - 63		16	20			
				8	10			
medium gravel	mittlerer Kies	6.3 - 20		4				
fine gravel	feiner Kies	2.0 - 6.3	2 - 60	2				
coarse sand	grober Sand			1				
		0.63 - 2.0	0.6 - 2.0	0.5				
medium sand	mittlerer Sand	0.2 - 0.63	0.2 - 0.6	0.25				
fine sand	feiner Sand			0.125				
		0.063 - 0.2	0.06 - 0.2	0.063				
coarse silt	grober Silt	0.02 - 0.063	0.02 - 0.06	0.02				
medium silt	mittlerer Silt	0.0063 - 0.02	0.006 - 0.02	0.0063				
fine silt	feiner Silt	0.002 - 0.0063	0.002 - 0.006	0.002				
clay	Ton	< 0.002	< 0.002	< 0.002				

Table 3.2 Necessary sample size for volume samples.

$Volume_{Sample}(m^3) = 2.5\,D_{max}\,(m)$	Fehr (1987a, b)
$Mass_{Sample}(kg) = 0.1\cdot10^b\,\rho_s\,D_{max}^3\,(m)$	Bunte & Abt (2001)

ρ_s = sediment density; accuracy: high (b = 5), medium (b = 4), low (b = 3).

Basically, a preliminary fractioning of the coarse components in the terrain is possible. This involves very coarse material with a grain size $D > 32$ mm that is measured individually (b-axis), grading these components according to the corresponding grain sizes and directly weighing them using weighing scales or otherwise by means of water displacement. That is, the weight is determined from the displaced volume of water and the unit weight (or specific gravity) of the grains. However, the influence of a subjective selection using a shovel is considerable. A possibility here would be taking a sample using an excavator shovel, and carrying out a complete measurement of the whole sample. Below a grain size $D < 32$ mm sieving should be used. If an armor layer is present, then, ideally, a sample of the armor layer should be taken together with a sample of the underlying layer. The thickness of the armor layer may then be assumed to be approximately D_{90} to D_{max}.

3.3.2　Analysis of surface bed material

The grain size distribution (GSD) of the surface bed material is measured with the aid of different methods of random sampling (grid, line, purely random), allowing measurements to be carried out for the surface of the bed or a gravel bank or a sediment exposure (BUNTE & ABT 2001). New methods of surface analysis are based on photographs of the streambed. With the so-called photo-sieving technique the limiting parameter is the grain size, since the projected area of the largest grain should take up at most 1% of the area of analysis. As a result, the apparent b-axis or the mean diameter of the exposed grain surface is determined by an automatic demarcation of the grain boundary. For this purpose there are already various programmed procedures that can be resorted to (WARRICK et al., 2009; BUSCOMBE et al., 2010; GRAHAM et al., 2010; DETERT & WEITBERCHT 2012a, b). However, the use of photo-sieving in torrent channels is problematic due to the very large granular components (and the required vertical distance of the camera). An alternative method is to use cameras capable of spatial, high-resolution distance measurement (NITSCHE et al., 2010).

In order to obtain the whole grain size distribution, the results of surface sampling methods (typically involving a minimum cut-off grain size) have to be converted into a standard recording method and combined with statistically-determined relationships for the finer material. These functional relationships depend on the grain shape, macrostructure, grading and porosity and are, therefore, only representative for some conditions of deposition. In mountain rivers and torrents of the Alps, the line by number analysis (LNA) developed at the Laboratory of Hydraulics, Hydrology and Glaciology, ETH, Zurich, is often used (ANASTASI 1984, FEHR 1987a, b). The method has mainly been tested in mountain rivers.

3.3.3　The Line by Number Analysis (LNA)
of the surface layer in streams with
coarse bed material

In streams with coarse bed material, sieve analysis of the surface layer often cannot be carried out, since either the sieve set for large grain diameters is unsuitable or the amount of bedload cannot be handled economically (see also Table 3.2). Therefore, simpler methods have been developed that take into account the conditions in coarse bedload streams. The most frequently used techniques for analyzing armor layers involve taking samples from a specific area using grids and lines (for an overview see BUNTE & ABT 2001). Here, the line by number sampling method LNA (FEHR 1987a, b) is explained in more detail.

3.3.3.1 Procedure for the execution of a line sample

A piece of string is stretched over the surface layer to be analyzed, helpingto avoid systematic errors in the choice of the particles to be investigated. For all particles that are greater than 1–2 cm and lie under the string, the middle axis (b-axis) is measured. The particles are divided into diameter classes (fractions) and counted. The line samples should include least 150 particles. If, at the same time, a volume-weight analysis of the fine material of the subsurface layer is carried out, then the classes of the line sample in the overlapping region should be adjusted using a volume sample of the latter.

3.3.3.2 Analysis of the line sample

The result of an LNA is a frequency distribution of the coarse fraction (>1 cm) of the surface particles on the streambed. To compare and combine the result with other methods, the proportions as expressed in percentages of the LNA values have to be converted to equivalent weight fractions in a volume analysis. FEHR (1987a, b) developed the following empirical conversion formula (for hydraulically loaded samples):

$$\Delta p_i = \frac{\Delta q_i\, D_{mi}^{\,0.8}}{\Sigma \Delta q_i\, D_{mi}^{\,0.8}} \tag{3.1}$$

where Δp_i = (weight of fraction i)/(weight of whole sample), Δq_i = (number of stones in the fraction i)/(number of stones in the whole sample), D_{mi} = mean grain diameter of the fraction i. Since the fines are underestimated by the LNA method, the grain size distribution curve (or grading curve) still has to be adjusted. It is noted that the theoretical exponent in Eq. 3.1 is 1.0 (KELLERHALS & BRAY, 1971) rather than the experimentally-determined value of 0.8 (FEHR1987a, b). For the conversion of an LNA into the grading curve of the subsurface layer, after FEHR (1987a, b) a proportion of fines of 0.25 for the sediments <1 cm may be assumed:

$$p_{ic} = 0.25 + 0.75 \sum_{1}^{i} \Delta p_i \tag{3.2}$$

Conversion of LNA to GSD of subsurface layer

If a conversion of an LNA into the grading curve of the surface layer (armor layer) is to be carried out, a smaller proportion of sediments <1 cm should be assumed. According to a summary by RECKING (2013) based on measurements in 78 gravel-bed streams, this proportion varies between about 0.01 and 0.2 and is, on average, 0.11. Therefore, for a conversion of an LNA sample into a surface layer grading curve it may be assumed that:

$$p_{ic} = 0.11 + 0.89 \sum_1^i \Delta p_i \tag{3.3}$$

Conversion of LNA to GSD of subsurface layer

where p_{ic} = corrected cumulative frequency (relative proportion) of the grains with $D \leq D_i$ (D_i = grain diameter of the grain size class i). To complete the grading curve for the fraction <1cm, either a volume sample is used (sieve, sedimentation analysis) or a distribution after FULLER is assumed (FEHR 1987a, b). Based on river sediments and assuming an optimum packing density, FULLER & THOMPSON (1907) developed a single parametric synthetic distribution based on D_{max} of the grading curve. The associated FULLER curve describes the grain size distribution of well-graded fluvial fine sediments. According to the investigations of MEYER-PETER & MÜLLER (1948), the FULLER curve provides a good approximation for mountain rivers in Switzerland and in the alpine region. Thus, a theoretical grading curve for the fines may be calculated as follows:

$$p_{FUi} = \sqrt{D_i \Big/ D_{max}} \tag{3.4}$$

where p_{FUi} = cumulative frequency of the grains with $D \leq D_i$ according to the FULLER curve. This grading curve after FULLER is then merged with the grading curve for the coarse part obtained using the LNA, where D_{max} is the maximum grain diameter of the fine sediments.

3.3.3.3   Statistical methods for the combination procedure

To combine a surface analysis with a volume analysis, the approach of ANASTASI (1984) was calibrated and verified by FEHR (1987a) for mountain rivers. FEHR (1987b) also recommended the rigid or flexible adjustment of the synthetic volume distributions after FULLER and subsequent combinationwith the converted surface distribution obtained from the LNA.

In the determination of characteristic grain sizes, a relatively large variability of the resulting grain size distribution must be expected due to the broad grain size distribution and the poor grading. Statistical methods for the LNA or similar analysis methods involve some uncertainty when used for torrents, since they were developed primarily for mountain rivers and have to be tested for torrent conditions. Nevertheless, it is usually found that, in the case of coarse torrent beds, the important characteristic grain sizes (D_{50}, D_{84}, D_{90}) are determined using surface analysis and are only moderately influenced by the uncertain combination with a sieve analysis.

3.4   BEDLOAD TRANSPORT IN STEEP CHANNELS

To understand bedload transport processes in mountain rivers and torrents requires consideration of several important factors, including the spatially-variable mobilization of the solids, the active sediment input, the formation or the mobilization of an armor layer or other stable morphological structures, sediment availability, and the transitional regime with debris-flow like sediment transport. It follows that there is not necessarily a functional relationship between channel discharge and bedload transport, the latter of which can vary considerably in a given channel as shown in Fig. 3.5 and Fig. 3.6.

To determine bedload transport quantitatively, the following aspects must be considered in particular: (i) initiation of transport, (ii) bedload transport function, (iii) partitioning of the flow resistance (additional energy losses), (iv) possible armor layer, (v) sediment availability.

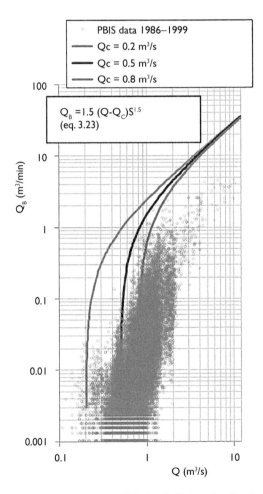

Figure 3.5 Bedload transport measurements in Erlenbach (Switzerland) using piezoelectric bedload impact sensors (PBIS) (RICKENMANN & MCARDELL 2007, eq. 7), and comparison with a laboratory-based bedload transport formula (Eq. 2.23) calculated for a channel slope $S = 0.105$.

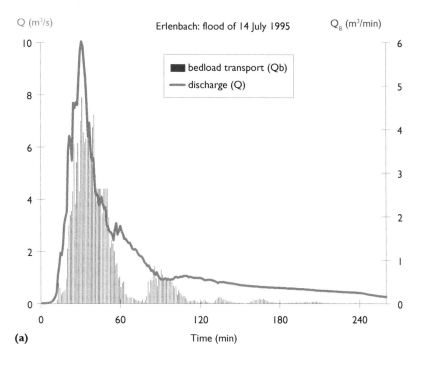

(a)

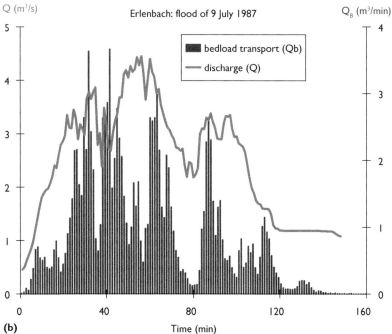

(b)

Figure 3.6 Bedload transport measurements in Erlenbach (Switzerland) using piezoelectric bedload impact sensors PBIS (Rickenmann & McArdell 2007, eq. 7), using examples (a), (b) from two flood events.

3.4.1   Fluvial bedload transport

In the case of fluvial bedload transport, the solid components are moved by the medium water. The material does not move continuously but intermittently with phases of inactivity, the grains sliding, rolling, or saltating, with little transport in the bottom layer and with intensive transport but decreasing concentration in the upper layers (SCHMIDT & ERGENZINGER 1992; SMART & JÄGGI 1983; RICKENMANN 1990). With increasing discharge, larger grains are moved; thus, after the start of motion, so-called selective bedload transport prevails (Fig. 3.7). Due to the selective bedload transport, mountain rivers and torrents can develop an armor layer. In the process the bed becomes coarse through washing out of the fines, and coarse components can nestle themselves into the bed, resulting in a higher stability with regard to renewed erosion. The bed structures thereby formed are an important feature of steep gravel bed channels (MONTGOMERY & BUFFINGTON 1997; ROSPORT 1998; WOHL 2000; SCHÄCHLI 1991; GRANT et al., 1990).

In torrents, bedload transport occurs often in the form of pulses or sediment waves (Fig. 3.6). This behavior can be traced back to spatially- and temporally-variable hydraulic conditions during the sediment transport or to discontinuous sediment availability and mobilization mechanisms that may be caused by turbulence peaks or retrogressive bed erosion after the scouring of coarse components. Hereby, the following transport processes are distinguished:

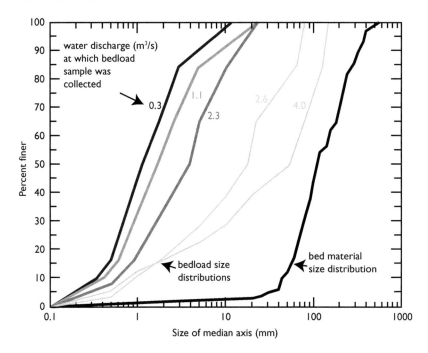

Figure 3.7 Composition of the sieve samples of the transported bedload for different discharges and stream bed material. The coarsening of the grain size distribution with increasing discharge is the result of selective bedload transport. The measurements are from the Roaring River (Colorado, USA), and the bed slope at the measuring section was approximately 5%. Modified from BATHURST (1987).

- Transport over an intact armor layer.
- Transport over an armor layer with limited sediment exchange.
- Transport with permanent sediment exchange between the transported solids and the bed sediment after breaking up the armor layer (ROSPORT 1998, HUNZIKER 1995, GÜNTER 1971).

Whereas at the beginning there is a selective, grain-size-dependent mobility of the solid material, all grains are essentially moving (equal mobility) after the complete break-up of the armor layer. This greatly affects the transported material in the grain size spectrum and it must be taken into account in the case of long-term transport simulations in which the first two transport processes play an important role (WEBB et al., 1997; HUNZIKER 1995).

A further parameter that is difficult to measure is the dynamic change of the flow resistance during the transport process. The transitions between grain, form and channel roughness are smooth. At the start of the mobilization of the bed, moreover, the surface structure changes and, therefore, the grain and form roughness also (ERGENZINGER & SCHMIDT 1995; ERGENZINGER et al., 1994). Characteristic grain size parameters (e.g. D_{30}, D_{50}, D_{90}) in mountain streams often exhibit fluctuation ranges of up to 30% (REID & DUNNE 1996; JÄGGI 1992). In particular, the temporal variability is difficult to take into account and causes a corresponding uncertainty in the analysis.

The solids transport can be expressed as a functional relationship between the following different independent parameters controlling the process:

- The critical shear stress or the critical discharge at the initiation of mobilization. These parameters are mainly relevant immediately after the start of bedload transport when the critical values are exceeded. It is important in the evaluation of the dynamics of the armor layer and also in the case of the selective transport of only a few sub-fractions of the entire spectrum of grain sizes present on the bed.
- The bedload transport capacity of the channel flow. After the onset of general bedload transport (over the whole grain size distribution) this parameter predominates. Bedload transport capacity is primarily expressed by the parameters hydraulic radius R or flow depth h, and channel slope S, or in simple approaches the first two may be replaced by the (unit) discharge q.
- Correction factor or calculation procedure to take into account the additional energy losses due to high flow resistance.

Natural channel beds of mountain rivers and torrents are typically characterized by a broad grain size spectrum. These conditions favor selective bedload transport during the rising or falling flood hydrograph, just after the initiation of mobilization or before the end of mobilization, and cause separation processes that can lead to the formation of an armor layer. For increasing discharge this selective bedload transport is followed by general bedload transport, during which the whole grain size spectrum is mobilized and transported due to stronger hydraulic forces (Fig. 3.8).

The range of weaker bedload transport between the initial mobilization and the general start of mobilization (discharge $Q_c < Q < Q_D$ in Fig. 3.8) is also described in the literature as phase-1 transport and the range with increasing bedload transport for $Q > Q_D$ as phase-2 transport. Phase-1 transport also corresponds to the transport of

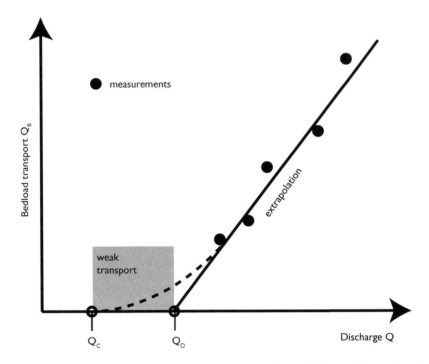

Figure 3.8 Increase of the bedload transport with discharge. In the figure Q_c denotes the initial mobilization, Q_D the general start of mobilization after the breaking up of the armor layer, implying the mobilization of the coarser grains of the surface layer. Modified from BEZZOLA (2005).

fines over a coarse armor layer. A thorough discussion of measurements and analyses on transport in phases 1 and 2 can be found in JACKSON & BESCHTA (1982), RYAN et al. (2005) and BATHURST (2007).

3.4.2   Start of mobilization

The start of mobilization is a key factor for the evaluation of bedload transport processes. At the beginning the mobilization is more of a random process in which individual grains separate and move from the bed matrix due to the action of shear stress peaks near the bed caused by turbulence (ROSPORT 1998). Based on investigations of the equilibrium stability between a flow-induced action (shear stress, form resistance, hydrodynamic uplift) and the resisting force of the grain (friction, tilting action, self-weight) in a sloping channel, SHIELDS (1936) formulated the concept of the dimensionless SHIELDS number θ. This number depends on the bed shear stress τ, the grain diameter D, the channel slope S, and the ratio of the sediment density to the density of water $s = \rho_s/\rho$.

$$\tau = \rho g h S \tag{3.5}$$

$$\theta = \frac{\tau}{g(\rho_s - \rho)D} = \frac{hS}{(\rho_s/\rho - 1)D} = \frac{hS}{(s-1)D} \tag{3.6}$$

The critical SHIELDS number at the start of transport is denoted by θ_c. Many bedload transport formulae are based on this assumption of a constant limiting shear stress for a given grain size, even if the spatial shear stress distribution near to the bed should be considered as a random parameter in a more exact analysis due to the effects of turbulence. With increasing channel slope, the erosional resistance of grains on the bed is theoretically reduced due to the slope-parallel weight component favoring destabilization (CHIEW & PARKER 1994). However, this effect is superimposed by reduced flow forces on the bed in the case of small relative flow depths at steeper channel slopes. Measurements show that the second effect predominates; that is, increasing values of θ_c were also determined for increasing channel slopes (Eq. 3.7, Eq. 3.8; Fig. 3.9).

$$\theta_c = 0.15\, S^{0.25} \tag{3.7}$$
$$\text{(LAMB et al., 2008)}$$

$$\theta_{ci} = (1.32S + 0.037)\left(\frac{D_i}{D_{50}}\right)^{-\gamma} \tag{3.8}$$
$$\text{(RECKING 2009)}$$

For a natural bed with a broad grain size distribution, the calculation of the bedload transport can also be carried out separately for the different grain fractions (classes of grain size). For this calculation of fractional bedload transport rates, a so-called hiding function is often used to modify the SHIELDS criterion compared to simpler calculations for a standard representative grain size only (PARKER 2008). In Eq. 3.8 the dependence on both the slope and also the hiding function are taken into account for the initiation of particle motion; the index i refers to a percentile of the grain size distribution. The exponent γ lies typically in the range of 0.64 to 1.0 (RECKING 2009). An exponent $\gamma = 1$ signifies that all grain sizes start moving at the same absolute critical bed shear stress, while with an exponent $\gamma = 0$ the absolute

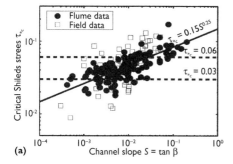
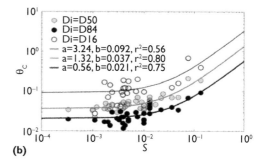

Figure 3.9 Variation of the critical SHIELDS number θ_c with channel slope S and empirical equations. (a) data from LAMB et al. (2008) with Eq. 3.7; where $\tau_{*c} = \theta_c$; (b) data from RECKING (2009) with lines obtained using the function $\theta_c = aS + b$. Eq. 3.8 with the exponents $\gamma = 0.93$ shows good agreement of $+/-50\%$ with the data of Fig. 3.9(b). Both data sets include flume and field data. Figure (a) from LAMB et al. (2008), Figure (b) from RECKING (2009), both with permission from Wiley/American Geophysical Union.

critical bed shear stress varies linearly with grain diameter (i.e. the SHIELDS number is constant and independent of the grain size to be moved).

In mountain rivers and torrents the shape of the bed, the low relative flow depth and the broad grain size spectrum exert a significant influence on the bedload mobilization (PALT 2001). For these channels, therefore, the following aspects should be considered with regard to the bedload mobilization:

- Ability to determine a representative grain size distribution, as well as the grain packing and the compactness of the grain structure
- Type of dependence between channel slope, characteristic grain size and relative flow depth
- Selective start of transport for different grain fractions
- Start of transport with or without armor layer
- Influence of the bed and channel geometry (deformed channel bed—plane bed after mobilizing all bed shapes)

For torrents, it is difficult to differentiate between the armor layer and the underlying sediment (subsurface layer), as a result of the heterogeneous origin of the bed sediments (heterogeneous mixture of various sediment sources). Thus, in a torrent, a grain size analysis of the coarse surface layer along the stream channel may reflect more the spatially-different lateral sediment inputs of scree material rather than the hydraulic sorting process. Due to the large grain size spectrum, determining complete grain size distributions is only possible using a statistical combination of different analysis methods (FEHR 1987a).

Many empirical investigations exhibit a relationship between channel slope, relative flow depth and start of mobilization. PALT (2001) traces the apparent relationship between channel slope and relative flow depth back to higher flow velocities at steeper slopes. Smaller relative flow depths are associated with a reduction of the near-bed flow velocity and, thus, they result in smaller shear stresses near the bed (BEZZOLA 2002). The apparent increased resistance of steeper channels is explained by PALT (2001) by the deformation of the bed due to the formation of bed structures that are initiated above a slope of 0.01. In steep channels there may be also additional flow resistance caused by the presence of large immobile boulders (e.g. YAGER et al., 2007; NITSCHE et al., 2012a).

After SHIELDS (1936), the critical shear stress for hydraulically-rough beds has the constant value of approximately 0.05. For non-uniform grain size distributions, MEYER-PETER & MÜLLER (1948) calculated a critical SHIELDS number θ_c of 0.047, which was determined for a range of slopes up to 2.3% and relative flow depths > 10. For S smaller than approximately 2%, θ_c lies typically in the range 0.03 to 0.06, for $D = D_{50}$ (BUFFINGTON & MONTGOMERY 1997). BEZZOLA (2002) considered the influence of grain shape and estimated the associated range of variation of θ_c to be about 40%, whereby θ_c would lie between 0.028 and 0.066 (c.f. also Fig. 3.9).

Since the determination of a representative mean flow depth is often difficult in torrents, the use of an alternative mobilization criterion is attractive, whereby a critical discharge per unit channel width, q_c, is determined instead of a critical SHIELDS number. Based on flume measurements for channel slopes in the range

$0.025 \leq S \leq 0.20$, BATHURST et al. (1987) proposed an approach (Eq. 3.9), which has been slightly modified by RICKENMANN (1991), namely (Eq. 3.10):

$$q_c = 0.15 \ g^{0.5} \ D_{50}^{1.5} \ S^{-1.12}$$

(3.9)
(BATHURST et al., 1987)

$$q_c = 0.065 \ (s-1)^{1.67} \ g^{0.5} \ D_{50}^{1.5} \ S^{-1.12}$$

(3.10)
(RICKENMANN 1991)

A more recent empirical equation by BATHURST (2013) (Eq. 3.11; Fig. 3.10) is also based on flume measurements and differs insignificantly from the earlier Eq. 3.9:

$$q_c = 0.13 \ g^{0.5} \ D_{50}^{1.5} \ S^{-1.146}$$

(3.11)
(BATHURST 2013)

An armor layer may be present in mountain rivers if the fine bed material is washed out during the receding limb of a flood hydrograph. For the critical discharge at the break-up of the armor layer (i.e. the start of transport of grains from the sub-surface layer) JÄGGI (1992), based on the investigations of GÜNTER (1971), proposed the following relationship for the dimensionless bed shear stress $\theta_{c,D}$:

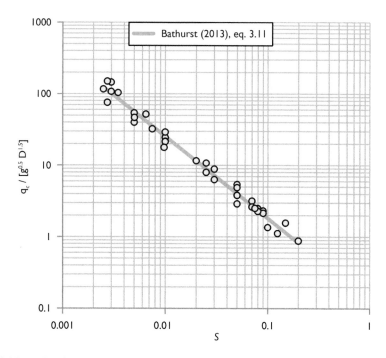

Figure 3.10 Normalized critical unit discharge q_c at initiation of bedload motion as a function of the channel slope S. Shown are (i) data from flume experiments performed with relatively uniform grain sizes, as compiled by BATHURST (2013) and (ii) Eq. 3.11 developed in the same study. Modified from BATHURST (2013).

$$\theta_{c,D} = \theta_c \left[\frac{D_{m,D}}{D_m} \right]^{2/3} \approx \theta_c \left[\frac{D_{90}}{D_m} \right]^{2/3}$$

(3.12)

(JÄGGI 1992)

where $D_{m,D}$ is the mean grain diameter in the armor (or surface) layer and D_m is the mean grain diameter of the subsurface layer; $D_{m,D}$ can be approximated and replaced by the D_{90} value of the subsurface layer (JÄGGI 1992). Using the Manning-Strickler equation, it can be shown that the discharge per unit width has the following proportionality: $q \sim h^{5/3} \sim \theta^{5/3}$. Thus, based on Eq. 3.12, the critical discharge per unit width at the break-up of the armor layer $q_{c,D}$ can be expressed as follows (where q_c may be determined using Eq. 3.9 or Eq. 3.11 (BADOUX & RICKENMANN 2008)):

$$q_{c,D} = q_c \left[\left(\frac{D_{90}}{D_m} \right)^{2/3} \right]^{5/3} = q_c \left[\frac{D_{90}}{D_m} \right]^{10/9}$$

(3.13)

(BADOUX & RICKENMANN 2008)

Making a similar transformation for steeper channels, it is preferable to base the critical discharge on the (simplified) VPE Eq. 2.17 for small relative flow depths, where the discharge per unit width is $q \sim h^{5/2} \sim \theta^{5/2}$. The critical discharge per unit width at the break-up of the armor layer $q_{c,D}$ can then be expressed as follows:

$$q_{c,D} = q_c \left[\left(\frac{D_{90}}{D_m} \right)^{2/3} \right]^{5/2} = q_c \left[\frac{D_{90}}{D_m} \right]^{5/3}$$

(3.14)

If it is assumed that, on average, $D_{90}/D_m \approx 2$, then, using Eq. 3.13, the ratio $(q_{c,D}/q_c)$ is given by $(q_{c,D}/q_c) = 2.2$ and with Eq. 3.14 the ratio is $(q_{c,D}/q_c) = 3.2$; that is, there is a considerable increase of the critical discharge for the break-up of an armor layer for steeper channel slopes.

Further approaches to account for the formation of an armor layer are described in PORTO & GESSLER (1999), HUNZIKER & JAEGGI (2002) and WEICHERT & BEZZOLA (2002). Another equation to determine the critical discharge $q_{c,B}$ on the break-up up or destruction of a "pavement-type layer" consisting of coarse granular material is based on investigations of the stability of block ramps with large blocks by WITTHAKER & JÄGGI (1986) for ramp slopes in the range $0.05 \leq S \leq 0.25$:

$$q_{c,B} = 0.257 \, (s-1)^{0.5} \, g^{0.5} \, D_{65}^{1.5} \, S^{-1.17}$$

(3.15)

(WITTHAKER & JÄGGI 1986)

Here, the grain size corresponds to D_{65}, a "mean" diameter for the large blocks, and, in a torrent channel, this quantity could be approximated roughly by the grain size D_{90}. Investigations into bedload transport carried out for flood events in the Valais in 2000 showed that the start of transport in flatter channels with S smaller than approximately 5% is sometimes clearly overestimated by Eq. 3.15 (BADOUX & RICKENMANN 2008).

Field measurements for the start of transport were made in mountain rivers in the Himalayas by Palt (2001). These data are compared with discharge-based approaches in steep channels in Fig. 3.11. This comparison suggests that the wide (grain size-dependent) range of values for the start of bedload transport in steep channels can be approximated by assuming lower and upper limits for the start of mobilization. As a lower limit, Eq. 3.9 or Eq. 3.11 are proposed, after Bathurst (2013). As an upper limit (transition to the general mobilization of all fractions), Eq. 3.14 or Eq. 3.15 after Whittaker & Jäggi (1986), for example, could be used.

Another interesting comparison of these discharge-based approaches for steep channels can be made with data from laboratory or controlled field environments where rock riprap was subjected to overtopping flow conditions on steep slopes. In the compilation of Abt et al. (2013), a total of 96 experiments are reported in

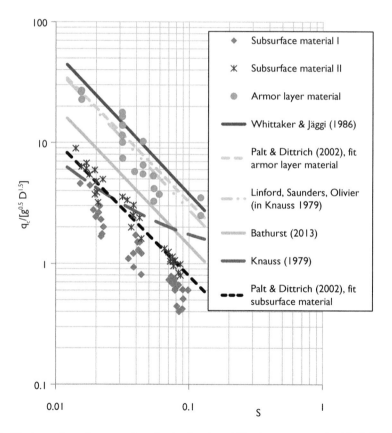

Figure 3.11 Discharge-based approaches for bedload mobilization compared with field data from mountain rivers in the Himalayas (Palt 2001), in the form of plots of normalized critical unit discharge q_c as a function of the channel slope S. The different equations proposed for q_c are given in Palt & Dittrich (2002) and in Abt et al. (2013). The field data refer to finer bedload material from subsurface layer transported over an armor layer, and to bedload material from the armor layer. Eq. 3.11 represents approximately a lower limit criterion, whereas Eq. 3.15 represents approximately an upper limit criterion.

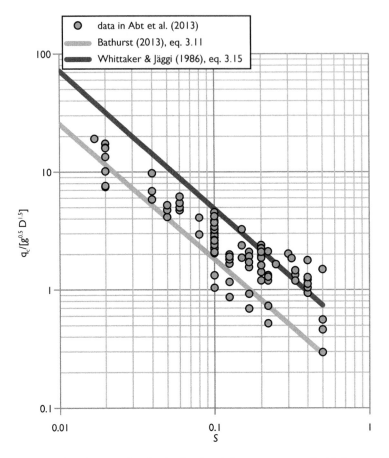

Figure 3.12 Discharge-based approaches for bedload mobilization compared with flume and field data for the failure of rock riprap layers at steep slopes (data reported in Aʙᴛ et al. (2013), shown as normalized critical unit discharge q_c as a function of the channel slope S. Eq. 3.11 represents approximately a lower limit criterion, whereas Eq. 3.15 represents approximately an upper limit criterion.

which discharge was increased until the riprap layer started to fail. Here, the failure unit discharge is set equal to the critical unit discharge q_c, and these data are compared with Eq. 3.11 and Eq. 3.15 in Fig. 3.12. The data from the riprap experiments tend to be closer to the upper limit criterion (Eq. 3.15) for the very steep slopes.

3.4.3 Approaches for calculating bedload transport

In the evaluation of bedload transport processes one must differentiate between approaches for calculating bedload transport capacity and approaches for calculating or estimating the probable actual bedload transport. The bedload transport capacity indicates a maximum transport level for given hydraulic conditions. The second type of approach should give a result closer to the actual bedload transport rates (e.g. taking account of additional energy losses), which for mountain rivers are up to almost an order of magnitude lower (JÄGGI 1992), and for torrents are up to several orders of magnitude lower (RICKENMANN 1997a, 2001a) than the transport capacity. The approaches to calculate the bedload transport capacity are based on flume experiments (often with a plane bed without armor layer) with an adequate sediment input from upstream, and, mostly, with relative flow depths greater than about 7 (MEYER-PETER & MÜLLER 1949; SMART & JÄGGI 1983; RICKENMANN 1990). Modified approaches were derived and tested, based generally on field data from channels with coarse stream bed structures, which partly included an armor layer and a limited sediment availability (PALT 2001; RICKENMANN 2005a; CHIARI et al., 2010; RICKENMANN & RECKING 2011; NITSCHE et al., 2011; SCHNEIDER et al., 2014).

The equations for calculating the bedload transport capacity can be presented in a standard way using the SHIELDS number θ (Eq. 3.6) and the dimensionless bedload transport rate Φ_b according to Eq. 3.16. The unit bedload transport rate q_b (per meter channel width) is then calculated using Eq. 3.17.

$$\Phi_b = \frac{q_b}{\sqrt{\left(\dfrac{\rho_s - \rho}{\rho}\right)gD^3}} = \frac{q_b}{\sqrt{(s-1)gD^3}} \qquad \text{(3.16)}$$

dimensionsless unit bedload transport rate

$$q_b = \Phi_b \sqrt{(s-1)gD^3} \qquad \text{(3.17)}$$

unit bedload transport rate

Based on flume investigations, MEYER-PETER & MÜLLER (1948) developed the well-known formula for bedload transport in gravel-bed streams, which is valid for channel slopes in the range $0.0004 \leq S \leq 0.023$:

$$\Phi_b = 8\left[\left(\frac{k_{st}}{k_o}\right)^{1.5}\theta - \theta_c\right]^{1.5} = 8\left[\theta' - \theta_c\right]^{1.5} \qquad \text{(3.18)}$$

MEYER-PETER & MÜLLER (1948)

The expression $(k_{st}/k_o)^{1.5}$ in Eq. 3.18 reduces θ to θ' when taking into account the energy losses due to form roughness. In Eqs. 3.6, 3.16 and 3.17, $D = D_m$ is to be used to determine θ and Φ_b, where D_m is the arithmetic mean value of the grain size distribution. After HUNZIKER (1995), the bed resistance was underestimated in these investigations and he proposed a reduction of the coefficient from 8 to 5 (see also HUNZIKER & JÄGGI 2002). A similar reduction of the coefficient was also proposed by WONG & PARKER (2006).

SMART & JÄGGI (1983) extended the investigations of MEYER-PETER & MÜLLER (1948) to steep channel slopes and developed a slightly modified equation valid for $0.0004 \leq S \leq 0.20$:

$$q_b = 4(s-1)^{-1}\left(\frac{D_{90}}{D_{30}}\right)^{0.2} qS^{1.6}\left(1 - \frac{\theta_c(s-1)D_m}{hS}\right) \qquad \text{(3.19)}$$

SMART & JÄGGI (1983)

Here, the ratio (D_{90}/D_{30}) represents an empirical correction of the transport efficiency related to the width of the grain size distribution, where, according to the experiments, the maximum ratio (D_{90}/D_{30}) was 10. This ratio is often exceeded in torrents and is subject to large fluctuations. The correction factor increases the bedload transport and is qualitatively in agreement with the increased bedload transport of the gravel fractions and larger sediments with increasing sand content in the surface layer, as proposed by WILCOCK & CROWE (2003).

RICKENMANN (1990, 1991, 2001a) analyzed the data of MEYER-PETER & MÜLLER (1948) and SMART & JÄGGI (1983), together with data from further flume tests taking into account increased concentrations of fine material in a clay suspension. He proposed the following equation to account properly for the effect of a changed fluid density:

$$\Phi_b = 3.1(s-1)^{-0.5}\left(\frac{D_{90}}{D_{30}}\right)^{0.2}\theta^{0.5}\left(\theta-\theta_c\right)Fr^{1.1} \tag{3.20}$$
$$\text{RICKENMANN (1990, 1991)}$$

Eq. 3.20 is valid for channel slopes in the range $0.0004 \leq S \leq 0.20$, and $Fr = V/(gh)^{0.5}$ is the FROUDE number. Eq. 3.20 was simplified by RICKENMANN (2001a) to Eq. 3.21 by approximating the exponent of Fr as 1.0, setting $(D_{90}/D_{30})^{0.2} = 1.05$ as for uniform bed material after SMART & JÄGGI (1983), and taking $s = 2.68$ for the density ratio of quartz sediment to water. For the development of Eq. 3.20, $D = D_m$ was used in Eq. 3.6 to determine θ and Φ_b, whereas, in later applications of Eq. 3.20 and Eq. 3.21, $D = D_{50}$ was used (NITSCHE et al., 2011; HEIMANN et al., 2015b). In addition, using the definitions for Φ_b and θ as well as the continuity equation $q = Vh$, Eq. 3.21 was transformed into the discharge-based equation Eq. 3.23 (RICKENMANN 2001a).

$$\Phi_b = 2.5\,\theta^{0.5}\left(\theta-\theta_c\right)Fr \tag{3.21}$$
$$\text{RICKENMANN (2001a)}$$

$$q_b = 3.1(s-1)^{-1.5}\left(\frac{D_{90}}{D_{30}}\right)^{0.2}\left(q-q_c\right)S^{1.5} \tag{3.22}$$
$$\text{RICKENMANN (2001a)}$$

$$q_b = 1.5\left(q-q_c\right)S^{1.5} \tag{3.23}$$
$$\text{RICKENMANN (2001a)}$$

According to the mathematical transformation, the term q_c in Eq. 3.22 and Eq. 3.23 has to be multiplied by V/V_c (with the critical flow velocity V_c, corresponding to the discharge for θ_c). However, this is neglected, since for q_c, mostly empirical functions are used. Eq. 3.22 is valid for channel slopes in the range $0.0004 \leq S \leq 0.10$; for higher slopes in the flume tests, due to larger bedload concentrations, the flow depth was significantly increased, which, in Eq. 3.20, is taken into account by θ but is not considered in Eq. 3.22. Eq. 3.22 has the advantage that, even without detailed information on the flow hydraulics, a comparison with field measurements is possible, provided that the discharge is known or can be estimated. The comparison of some formulae with the flume data of the hydraulics laboratory at ETH Zurich (VAW-ETH) is shown in Fig. 3.13.

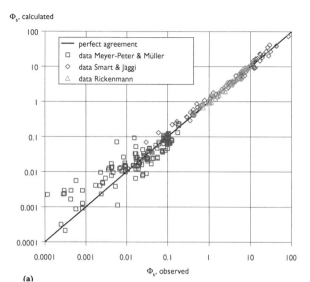

(a)

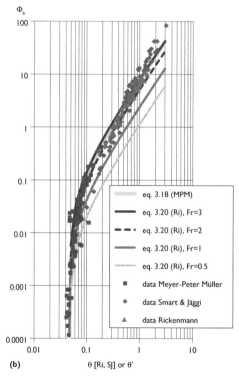

(b)

Figure 3.13 (a) Comparison of the transport rates measured in laboratory flume tests (VAW-ETH data) with the transport rates calculated using Eq. 3.20, expressed as dimensionless transport rate. (b) Comparison of Eq. 3.18 after MEYER-PETER & MÜLLER (1948) (MPM) and Eq. 3.20 after RICKENMANN (1990) (Ri) with the same flume data as in (a). In figure (b) the MPM data were corrected using θ' in Eq. 3.18.

For the flume data of VAW-ETH with channel slopes in the range $0.03 \leq S \leq 0.20$ the following equation was determined:

$$q_b = \frac{12.6}{(s-1)^{1.6}} \left(\frac{D_{90}}{D_{30}}\right)^{0.2} (q-q_c) S^2 \tag{3.24}$$
$$\text{RICKENMANN (1990, 1991)}$$

Eq. 3.24 shows better agreement with the data for $S \geq 0.10$ than Eq. 3.22. Other bedload transport investigations for steep channel slopes (MIZUYAMA 1981; WARD 1986) led to an equation similar to Eq. 3.24 with an exponent of 2 for the channel slope factor. Fig. 3.14a compares measured bedload transport rates and rates calculated with Eq. 3.24 using the VAW-ETH data for steep channels. In the same figure two other independent data sets are also shown: the flume tests of AZIZ & SCOTT (1989) in a conventional flume were obtained for channel slopes in the range $0.03 \leq S \leq 0.10$ and with sand grain sizes from 0.29 to 1 mm; the flume tests of NNADI & WILSON (1992) were carried out in a closed horizontal channel under pressure, with pressure gradients equivalent to $0.013 \leq S \leq 0.206$, sand grains of 0.7 mm size and nylon particles of 4 mm size.

For steep slopes, the weight component of sediment grains parallel to the channel slope contributes to the bedload transport, and ABRAHAMS et al. (2001) or ABRAHAMS (2003) based on SCHOKLITSCH (1914) proposed the following correction, which incorporates an increased slope factor S_k as follows:

$$S_k = S \left(\frac{\sin \phi_s}{\sin(\phi_s - \beta)}\right) = S a_k \tag{3.25}$$
$$\text{(ABRAHAMS 2003)}$$

where ϕ_s is the natural slope angle (friction angle) of the bedload particles under water and β is the angle of the channel slope. Eq. 3.25 in combination with Eq. 3.22 was applied to the VAW-ETH data as well as to those of AZIZ & SCOTT (1989). The results are shown in Fig. 3.14(a, b), together with the bedload transport rates for the data of NNADI & WILSON (1992), calculated using Eq. 3.22. The two different approaches lead to a similarly good agreement with the observed values. With the slope correction, Eq. 3.25, an equation of the type Eq. 3.22 or Eq. 3.23 can be applied over a very large range of slopes $0.0004 \leq S \leq 0.20$. At high transport rates, the mixture flow depth, as compared with the purely water-flow depth, is increased significantly, resulting in a large bed shear stress. This effect is considered implicitly in Eq. 3.20, since the mixture flow depth was used in its derivation.

3.4.4   Consideration of energy losses

The transport formulae for steep channel slopes, e.g. Eq. 3.20 or Eq. 3.22, can be extended using the efficiency factor α (i.e. by multiplying a "constant" coefficient of e.g. 1.5 in Eq. 3.23 by the factor α), which can adopt values between 0.001 and 1 (RICKENMANN 2001a). In mountain rivers and torrents with bed slopes

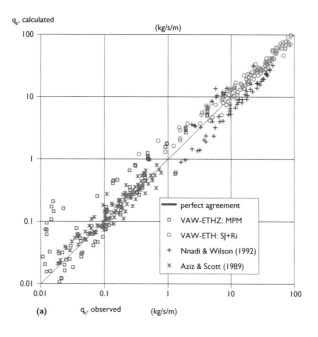

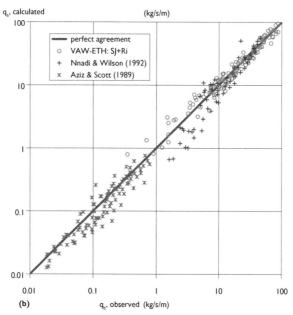

Figure 3.14 (a) Comparison of the measured bedload transport rates with the values calculated using Eq. 3.24 for the VAW-ETH data for steep channel slopes and two independent data sets from Aziz & Scott (1989) and Nnadi & Wilson (1992). (b) Comparison of the measured bedload transport rates with the values calculated using Eq. 3.22. For the VAW-ETH data (MPM: Meyer-Peter & Müller; SJ: Smart & Jäggi; Ri: Rickenmann) of all channel slopes and the data of Aziz & Scott (1989) the slope correction using Eq. 3.25 was taken into account (after Rickenmann 2005a).

steeper than about 3 to 5%, observed values of bedload transport were found, in general, to be smaller than values calculated with bedload transport equations (RICKENMANN & KOSCHNI 2010). Here the following aspects should be considered:

- The influence of form losses or macro-roughness effects on flow resistance is difficult to quantify.
- The start of substantial bedload transport is difficult to quantify, since, with a broad distribution of grain sizes in a river bed, the flow conditions and the start of movement are influenced in a complex manner.
- The bedload transport is often limited by the sediment availability and not by the transport capacity.
- For bed slopes steeper than about 10%, according to Eq. 3.20 or Eq. 3.24 high solids concentrations result, which are more plausible for debris floods (transitional regime) or debris flows than for fluvial bedload transport (RICKENMANN 2005a, 2012).

To determine bedload transport taking into account energy losses due to coarse roughness elements (so-called macro-roughness), Eq. 3.20 or Eq. 3.22 (or any other transport equation) can be applied in combination with a method for the partitioning of the flow resistance (see chapter 2.4). RICKENMANN (2005a) introduced an empirical function to estimate a reduced slope of the energy line for large-scale roughness conditions in steep channels. With this method a better agreement between bedload volumes observed in nature and calculated could be achieved for various flood events (e.g. CHIARI et al., 2010; CHIARI & RICKENMANN 2009, 2011; BADOUX & RICKENMANN 2008). This approach was modified by RICKENMANN & RECKING (2011), who used an extended database. The partitioning of flow resistance is based on an earlier proposal by MEYER-PETER & MÜLLER (1948) and later tested by PALT (2001). The reduced energy slope S_{red} is calculated with reference to a base level of the flow resistance (for a basic roughness of the bed material) and determines the energy that is available for the bedload transport:

$$S_{red} = S\left(\sqrt{\frac{f_o}{f_{tot}}}\right)^e = S\left(\frac{n_o}{n_{tot}}\right)^e$$

(3.26)
(RICKENMANN & RECKING 2011)

According to the r DARCY-WEISBACH flow law (Eq. 2.2), the slope of the energy line S is proportional to the friction coefficient f or, according to the flow law after MANNING-STRICKLER, Eq. 2.10, it is proportional to the Manning coefficient n squared, and the exponent e should have the value 2. MEYER-PETER & MÜLLER (1948), based on theoretical considerations, showed that e can also take on smaller values (down to 1.33), and based on their experimental results they proposed an empirically-determined value $e = 1.5$. RICKENMANN et al. (2006a) suggested that plausible values for e may lie in the range $1 \leq e \leq 2$.

For the calculation of bedload transport the reduced energy slope S_{red} is either introduced via θ in Eq. 3.20 or directly in Eq. 3.22. The values for θ_c are determined empirically and refer in general to the total flow resistance (or the total bed shear stress). Thus, in the use of Eq. 3.20 θ_c also has to be reduced. The reduced value of θ_c can be determined such that $\theta_{c,r} = h_c \, S_{red}(h_c) \, [(s-1) \, D_{50}]^{-1}$ corresponds to the discharge conditions at the start of bedload transport, i.e. $S_{red}(h_c)$, and thus $\theta_{c,r}$ is constant for a given channel slope and a given grain size distribution (D_{50}, D_{84}) (NITSCHE et al., 2011, 2012b). Alternatively, θ_c can also be reduced as follows, using a discharge-dependent value of S_{red}: $\theta_{c,r} = \theta_c \, (S_{red}/S)$. It is difficulty to verify which of the two approaches is more plausible.

The approach of RICKENMANN & RECKING (2011) given in chapter 2.4 for the partitioning of flow resistance is basically a function of the relative flow depth. However, as a general empirical approach, it implicitly contains information about a mean increase in roughness in steep and rough channels. In the study of NITSCHE et al. (2011), other ways of partitioning the flow resistance were investigated, including, for example, consideration of the additional energy losses caused by large immobile boulders (YAGER et al., 2007; WHITTAKER et al., 1988) or by step-pool sequences (EGASHIRA & ASHIDA 1991). All these approaches were combined with Eq. 3.21 and Eq. 3.26 with an exponent $e = 1.5$, and the calculated bedload transports were compared with observations of the transported bedload volumes (flood events in Switzerland in 2005; flood events in canton Valais, Switzerland, in 2000; long-term discharge and bedload measurements of the WSL institute in torrents in Switzerland). Overall, for all channel types (stream bed morphologies), the best results were obtained with the empirical approach of RICKENMANN & RECKING (2011) and with the more physically-based approach of YAGER et al. (2007). A summary of these results is presented in Fig. 3.15.

Table 3.3 shows the combinations of equations used for the bedload transport calculation and for the partitioning of the flow resistance, according to which the results are arranged in Fig. 3.15. For detailed investigations for a given channel type (e.g. influence of large boulders in different concentrations), specific approaches should be preferred (YAGER et al., 2007; or WHITTAKER et al., 1988 for channel slopes not exceeding about $S \approx 0.07$); however, these approaches require more exact investigations of the riverbed morphology. To have some idea about the uncertainty of the estimates of bedload transport, the most suitable approaches for partitioning the flow resistance for a given channel type can be used to examine the range of possible results.

The exponent $e = 1.5$ used in NITSCHE et al. (2011, 2012b) and in SCHNEIDER et al. (2014) is near the range of the best exponents according to simulations with the software SETRAC using an earlier approach for flow resistance partitioning (CHIARI & RICKENMANN 2009, 2011).

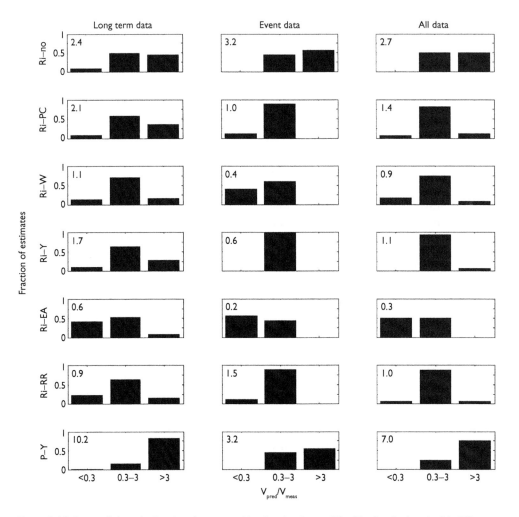

Figure 3.15 Ratio of the calculated and measured bedload volumes (V_{pred}/V_{meas}), calculated with different combinations of equations (rows; defined in Table 3.3) and differentiated according to data groups. The (V_{pred}/V_{meas}) ratios are presented in three classes, of which the middle class represents all calculations within a factor 10 of the measured bedload transports. The gray numbers indicate the mean value of the (V_{pred}/V_{meas}) ratios. The group «Long-term Data» consists of 207 transport events, while the group «Event Data» consists of 9 transport events. The group «All Data» consists of the summed bedloads of the individual channels, in order to weight each stream independently of the number of events in the same way. The approach of WHITTAKER et al. (Ri-W) was not used for 4 channels. From NITSCHE et al., 2011, with permission from Wiley/ American Geophysical Union.

Table 3.3 Abbreviations used for the combination of equations for bedload transport and for the partitioning of flow resistance leading to the results shown in Fig. 3.15.

Bedload transport	Equation	Partitioning of flow resistance	Equation	Abbreviation
RICKENMANN (2001a)	Eq. (22)	No reduction	–	Ri-no
RICKENMANN (2001a)	Eq. (27)	PAGLIARA & CHIAVACCINI (2006)	Eq. (10) + (11)	Ri-PC
RICKENMANN (2001a)	Eq. (27)	WHITTAKER et al. (1988)	Eq. (3) + (4)	Ri-W
RICKENMANN (2001a)	Eq. (27)	EGASHIRA & ASHIDA (1991)	Eq. (7) + (9)	Ri-EA
RICKENMANN (2001a)	Eq. (27)	YAGER et al. (2007)	Eq. (13) + (14)	Ri-Y
RICKENMANN (2001a)	Eq. (27)	RICKENMANN & RECKING (2011)	Eq. (15) + (16) + (17)	Ri-RR

3.4.5 Transition to debris flood and debris flow

The mechanical behavior of debris flows is complex and depends on different factors such as viscosity and turbulence of the mixture, dispersive forces due to collision of the coarse components, friction forces between the (coarser) grains and the shear strength of the matrix (consisting of fine components and water). Depending on the dominance of these factors, a rough classification into mudflows and (granular) debris flows is possible. This simplified presentation does not, however, take into account the property that the solids concentration often decreases in the rearward region of a mudflow or debris flow surge. Debris-flood conditions of solids transport can be caused by the sudden input of solids, by the liquefaction of larger streambed reaches, or after breaking through logjams and dams (TAKAHASHI 1991).

For a volume proportion of more than about 5% silt or clay, debris-floods or hyper-concentrated flows exhibit an increasing viscous behavior. From a volume proportion of solids in total of 45 to 55% a debris flow or mudflow develops (COSTA 1984; JULIEN & O'BRIEN 1997). Empirical investigations in Switzerland indicate that debris flows initiate typically in steep scree with slopes between 40 and 58%, in the contact zone between bedrock/scree with slopes between 45 and 70%, in gullies with slopes between 45 and 70% and in stream channels with slopes between 23 and 65% (HAEBERLI et al., 1991; RICKENMANN & ZIMMERMANN 1993). These results are in agreement with flume investigations with uniform sediments for which a change can be observed from the usual mobilization mechanism to a sliding-type en mass instability of the channel bed for slopes larger than 20% (SMART & JÄGGI 1983). Thus in torrent channels with bed slopes of more than ca. 20% and in the absence of stabilizing bed structures, transport processes with the characteristics of debris flows can be expected (JÄGGI 1992). The transport formulae of SMART & JÄGGI (1983) and RICKENMANN (1990) predict that for channel slopes steeper than about 10 to 15% and high flow intensities, solids concentrations occur that are typical for debris-flood or debris-flow conditions. In the case of high flow resistance (structured torrent channels) and discharges close to the start of transport such formulae tend to overestimate the observed transport rates considerably if no correction for the effective shear stress is made (such as discussed in chapter 3.4.4).

The flume experiments of SMART & JÄGGI (1983), RICKENMANN (1990) as well as similar laboratory and field observations indicate that very high sediment concentrations occur for channel slopes steeper than about 20%. An extrapolation to even steeper slopes leads to similarly high transport rates, as observed in field tests on the formation of debris flows. Further, above limiting slopes of approximately 20–25%, a general instability of the bed has to be considered, together with a continuous transition from fluvial bedload transport to debris floods and debris flows. Likewise, comparisons of discharge criteria for the initiation of bedload transport and for debris-flow formation (as well as of empirical flow resistance laws for water runoff in steep channels and debris flows) indicate that a continuous transition is to be expected (RICKENMANN 2012). A fairly continuous transition of transported sediment loads was observed between fluvial sediment transport and debris flows for a large rainstorm event in Switzerland (Fig. 3.16).

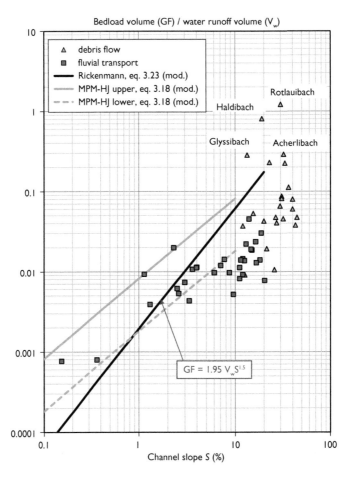

Figure 3.16 Data from different types of transport processes indicate a relatively continuous transition from fluvial transport to debris flows. The data comes from the flood events of August 2005 in Switzerland. The purple line corresponds to the integration of Eq. 3.23 for fluvial bedload transport over the flood period, wherein a pore volume (voids content) of the deposited material of 30% is considered, and the channel slope S was determined upstream of the deposited material. The modified Eq. 3.23 (MPM-HJ) refers to the equation of MEYER-PETER & MÜLLER (1948) but using a coefficient of 5 according to HUNZIKER & JÄGGI (2002), and also accounting for 30% pore volume. With the 4 debris-flow data points denoted by the names of the streams, much sediment entered the channel due to large landslides. Modified from RICKENMANN & KOSCHNI (2010).

3.4.6   Deposition slope behind check dams

The expected deposition slope upstream of check dams or in sediment retention basins is an important parameter for the design of protection measures. In designing a series of check dams, the deposition slope between successive check dams has to be known or assumed to determine the exact locations (spacing) and heights of the individual check dams. In the case of a sediment retention basin, the expected deposition slope essentially defines (together with the width of the basin) the sediment volume that can be stored to reduce the sediment load that will be transported further downstream during an event.

According to bedload transport equations, the deposition slope will increase with increasing sediment input from upstream and with coarsening of the transported solids. However, both the expected runoff hydrograph and the sediment input from upstream are difficult to determine. In addition, in steeper channels and headwater catchments, it is possible that not only fluvial bedload transport occurs but also debris flows. These factors all complicate a "theoretical" estimation of the deposition slope.

Therefore, in engineering practice, deposition slopes S_{dep} are mostly estimated based on experience. A frequently-taken assumption is that S_{dep} may vary in the range from $(1/2)S_o$ to $(2/3)S_o$, where S_o is the original stream bed slope (IKEYA 1979; ROMANG 2004; PLANAT 2008; PITON & RECKING 2015a). In Japanese design guidelines for the construction of sediment retention basins it is recommended to use the same range of expected deposition slopes (PWRI 1988). Published data on deposition slopes are very rare (e.g. ROMANG 2004); some observations made in steep streams in Italy (PORTO & GESSLER 1999) and in Iran (NAMEGHI et al., 2008) are illustrated in Fig. 3.17.

3.4.7   Comments on the estimation of the solids transport

In torrents, due to the very variable delivery of solid material, a very broad grain size distribution and a spatially-variable sediment availability, one must expect a very large fluctuation of the solid material transport rate and a very noticeable phase of selective transport. This implies that:

- The solids transport rate, especially for small to medium flow intensities, only has a limited functional relationship with the discharge. As an upper threshold, the calculated transport capacity (for quasi-plane bed conditions and unlimited sediment availability), has the highest relative accuracy. The lower range can be estimated from the (often limited) availability of solids, the use of an armor layer criterion or a higher critical shear stress at the start of mobilization, or from a consideration of energy losses.
- Owing to the broad grain size distribution and very heterogeneous sediment availability, selective bedload transport takes place for small to medium flow intensities. This favors the formation of an armor layer and results in spatially strongly variable grain size distributions. The macro-roughness elements of the bed reduce the effective shear stress acting near the stream bed.

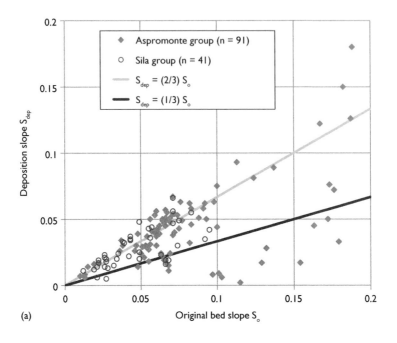

(a)

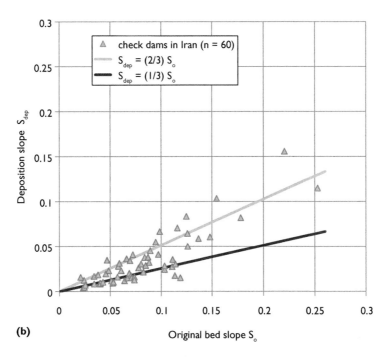

(b)

Figure 3.17 Deposition slopes behind a series of check dams in comparison with the original slope of the stream bed for (a) data from two Calabrian streams in Italy with gravel bed sediments with D_{50} ranging from about 5 to 20 mm (PORTO & GESSLER 1999), (b) data from a stream in Iran with mostly sandy sediments with D_{50} typically ranging from about 0.4 to 2 mm (NAMEGHI et al., 2008).

3.4.7.1 Critical bed shear stress or critical specific discharge

Besides the characteristic grain diameter and the channel slope, the basis for analyzing the bedload transport is primarily the ratio of actual to critical shear stress. For steep channels, the critical shear stress reacts directly and very sensitively to variations in flow depth, which is why some calculation methods also use the simpler approach of specifying a critical specific discharge. More recent investigations show the tendency of an increase in the critical dimensionless bed shear stress θ_c with increasing channel slopes (Lamb et al., 2008; Recking 2009; Bunte et al., 2013), where by θ_c is determined using the total bed shear stress.

3.4.7.2 Transport reduction effect of the armor layer—selective transport

Calculating θ_c or q_c according to the armor layer criteria may result in considerably reduced transport rates. This may be appropriate if no extreme events have to be considered, and when relatively small discharges occur over longer time periods. For the range of fluctuation of the start of transport, with or without an armor layer, a factor of about 2 to 3 with regard to θ_c or q_c can be expected.

3.4.7.3 Calculation of solids transport

In torrents with typically steep channel slopes, only a few transport formulae have been tested with field data. The two extreme cases consist of determining the transport capacity (maximum possible transport rate) and a reduced transport rate due to high flow resistances (considering additional energy losses, e.g. with Eq. 3.26). If bedload transports are calculated according to the transport capacity, then these conditions may be more representative for debris floods or debris flows. Calculations taking into account a reduced energy slope are valid for conditions with fluvial bedload transport.

3.5 DRIFTWOOD IN TORRENTS AND MOUNTAIN RIVERS

In mountainous and forested catchments wood can find its way into the stream channel through landslides, debris flows, erosion, snow avalanches or storms (windthrow) (Rudolf-Miklau et al., 2011). Depending on the type and origin, one speaks also of dead wood, old wood, fresh wood and wood from trees uprooted by avalanches. For water-related transport and in-channel deposits of logs and rootstocks, the term "large woody debris" was in use for some decades but has been replaced more recently by "large wood" (e.g. Wohl et al., 2010; Jackson & Wohl 2015).

3.5.1 Flood hazards associated with driftwood

An overview of the problem of floating woody debris in mountain rivers and torrents during high discharges is given by Hartlieb & Bezzola (2000), Mazzorana et al. (2009, 2011), Rudolf-Miklau et al. (2011), Comiti et al. (2012), and

RUIZ-VILLANUEVA et al. (2014). In flood events large pieces of wood often cause problems due to logjams at bridges, culverts or even natural constrictions like gorges. The most important effects are: (i) logjams or temporary blockages obstructing water flow and bedload transport in natural channel sections, which can favor the formation of debris flows in steep reaches, (ii) overtopping of water and sediment out of the channel onto the fan or banks can lead to large amounts of deposited material and debris. Another frequent and undesirable consequence of excessive large wood is the partial or complete clogging of open check dams of sediment retention basins, whereby the desired regulation effect (dosage) regarding bedload transport during a flood event is impaired or completely inhibited (PITON & RECKING 2015b). Further, it can also lead to the destruction of bridges, or large wood pieces can cause impact damage to buildings.

3.5.2   Origin and amount of large wood in stream channels

Information on the type and origin of the wood in the channels may be found in HARTLIEB & BEZZOLA (2000), RIMBÖCK (2003), HASSAN et al. (2005b), MAZZORANA et al. (2009), KASPRAK et al. (2012), and GURNELL (2013). Once the wood lies in the channel, the following aspects are important: the shape and the dimensions of the individual elements, whether the pieces have branches, and, particularly, the presence of and the proportion of rootstock. In addition, the type of wood and the water absorption, and thus the density of the wood, influence the degree of mobilization in the channel during floods.

Concerning possible amounts of wood, it is possible todifferentiate between the effective amount of driftwood that is transported during a flood and the amount of potential wood that can be mobilized from within the channel or supplied from areas near the river. Investigationsbased on data from the Swiss Alps, Japan and North America show that both the amount of transported wood debris and the wood debris potential can be correlated roughly with the catchment size (RICKENMANN 1997b; WALDNER et al., 2008). Furthermore, the transported amount of driftwood also depends on the integrated water runoff or on the transported bedload volume of a flood event (RICKENMANN 1997b). To make more exact statements about the potential amount of large wood, detailed investigations in a given catchment are necessary, whereby factors like state of the forest, erosion processes, channel profile and bed slope need to be considered. RIMBÖCK (2003) developed a method for estimating the wood debris potential based on aerial photographs.

For the floods of 2005 in Switzerland a budget of large wood was determined for selected catchments, including a quantitative assessment of wood recruitment processes, namely of landslides, debris flows, bank erosion, and entrainment of in-channel wood (WALDNER et al., 2007, 2008). The contribution of in-channel wood was based additionally on an earlier study of wood in torrent channels (RICKLI & BUCHER 2006). A GIS procedure was developed for estimating potential wood debris contributions due to landslides (MÄCHLER, 2009). Essentially, the new data confirm the approximate relationship between the amount of transported large wood in flood events as a function of catchment area (Fig. 3.18).

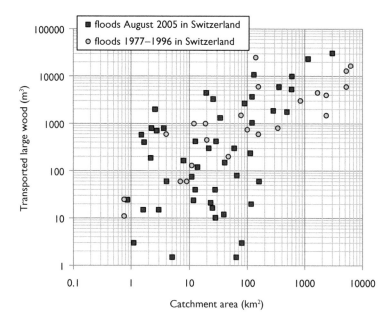

Figure 3.18 Transported volumes of large wood as a function of the catchment area, as observed after flood events (mainly in Switzerland) (WALDNER et al., 2008).

3.5.3 Transport of driftwood and logjam hazards

The initial transport of the pieces of wood lying in the channel depends mainly on the flow conditions, the type of wood (density, presence of branches, roots) as well as the location with respect to the flow action (BEZZOLA et al., 2002; BRAUDRICK & GRANT 2000, 2001). Flume investigations indicate that the transport of logs begins at relative flow depths $h/d \sim 0.5$ to 1.5, where d is the log diameter. The critical values of h/d increase in the above-mentioned range with increasing FROUDE number (as well as with the increasing number of branches or with the presence of rootstocks), whereas the critical values for h/d with simultaneous bedload transport are more likely to lie in the lower range.

The transport distances of pieces of wood increase if the lengths of logs are smaller than the mean channel width (LANGE & BEZZOLA 2006; SCHMOKER & HAGER 2011; LASSETTRE et al., 2012; LUCIA et al., 2015). Generally, driftwood floats at the surface. After RIMBÖCK (2003) coniferous wood (softwood), rootstocks and old dried wood usually float on the surface due to their low density, whereas the heavy oak and beech woods seldom float. Bulky, very branched wood, and rootstocks only float in the case of large flow depths, and, otherwise, are often transported by rolling over the stream bed. During the transport process the position of the wood continually changes due to turbulent flow. Often the wood is aligned parallel to the direction of flow. During transport the wood debris becomes smaller in size. ZOLLINGER (1983) reports that a whole tree with the crown and rootstock can have its branches removed and its bark peeled off, and that the tree may be broken into 1 to 5 m long pieces within relatively few meters during transport in a torrent channel.

The deposition of driftwood takes place under natural conditions as soon as the discharge decreases and the buoyancy and the force of flow for further transport are no longer sufficient. After deposition of individual pieces and with the accumulation of additional wood, fairly shallow heaps may be formed. With the transport of short individual logs there is only a relatively small risk of logjams, since the logs can easily align themselves in the flow direction and thus pass a constriction.

An overview of possible measures to reduce the risk of logjams is given by LANGE & BEZZOLA (2006). The risk of logjams near bridge cross-sections was investigated in flumes with hydraulic model tests (BEZZOLA et al., 2002). For a batch-wise delivery of wood mixtures the probability of logjams p_v (number of tests with logjams in relation to all tests of the same category) reached values of 0.2 to 1.0. Noticeably, p_v was clearly larger if rootstocks were present. The probability of logjams of individual pieces of wood depends mainly on their dimensions relative to the width of the critical cross-section. In the case of individual logs, an increase of up to p_v = ca. 0.4 was obtained in the range $0.5 < L_W/B < 2$, where L_W = length of individual log and B = width of the opening. In the case of rootstocks a striking increase of up to $p_v = 1.0$ was obtained in the range $0.6 < d_W*/H < = 1$, where H = clear height of the critical cross-section, $d_W* = (d_{Wmax} \, d_{Wmin} \, L_h)^{1/3}$, with d_{Wmax}, d_{Wmin} = maximum and minimum, respectively, of the dimension of the root plate and L_h = length of trunk extension. To reduce the risk of logjams it is recommended that the bed width of the channel should be about twice the dimension of the expected maximum length of the logs, and that the clear height under the bridge should be at least 1.7 times the critical dimension of the expected rootstocks. The tests also show that the amount of driftwood is primarily important for the temporal evolution of the logjam process. Whether logjams occur depends, in the first instance, on the dimension and shape of the largest components (LANGE & BEZZOLA 2006).

3.6   CRITICAL CHANNEL CROSS-SECTIONS AND POTENTIAL DEPOSITION

The assessment of the flood hazard along torrents and mountain rivers carrying bedload basically requires consideration of three questions: (i) Is the hydraulic conveyance capacity of the existing channel adequate to discharge the flood without damage?; (ii) When and where can intensive bedload transport lead to deposition with associated flow overtopping?; (iii) When and where can high flows with little bedload lead to erosion that could endanger the stability of banks and the foundations of structures? The assessment of these three aspects can be carried out at two different levels of detail: (a) with simple estimates of the hydraulic conditions and of the bedload transport at critical locations (cross-sections), combined with an integrative assessment of possible effects for the entire duration of the flood event; or (b) using numerical models to simulate the hydraulics and the bedload transport, though experience with these models is limited thus far for steeper slope ranges and for flows overtopping a channel on the fan (e.g. CHIARI & SCHEIDL 2015).

For the procedure with the simple estimates (a), the analysis methods presented in the previous sections can be applied. If significant sediment depositions occur in the channel area, flow overtopping is to be expected, together with deposition of bedload outside the channel (e.g. on the fan). Especially prone to critical depositions are sudden concave changes in the longitudinal profile (decrease in channel slope without increase of discharge). If there are also bridges at such critical locations, the risk of a complete blocking of the flow cross-section is especially high in the case of driftwood in addition to bedload transport. Analytical methods for predicting the depositional behavior for a sudden change of slope are described in BEZZOLA et al. (1996) and in FRENCH et al. (2001).

In calculating the bedload transport capacity of the channel on the fan, special attention is required in the case of an artificially-paved (or concreted) and smooth channel bed. In this case, the bedload transport capacity is considerably higher than in a natural channel with a movable bed and modified calculation approaches are necessary (HUNZINGER & ZARN 1996; SMART & JÄGGI 1983). If the bedload-carrying flow leaves the channel, it is necessary to predict the flow paths and areas of deposition on the fan. This can be done mainly based on the fan topography, but structures (buildings, roads) can also influence the flow and deposition behavior considerably. Therefore, especially in populated areas, different scenarios of the flooding process may have to be considered, depending on the depositional process (which can be influenced in addition by the amount of driftwood). Basically, in the case of a spreading out of fluvial deposition on the fan, the entire bedload volume (deducting the portion deposited in the channel) has to be distributed along the flow path. The average thicknesses of such deposits are likely to be less than for debris-flow deposits on the fan. In the case of fluvial transport, the coarser bedload grains tend to be deposited in steep zones, whereas the finer grains may be transported further downstream to flatter zones.

3.7 NUMERICAL SIMULATION MODELS

There are several sophisticated hydraulic-sedimentologic numerical simulation models for gravel-bed and sand-bed streams with limited channel slopes. For steep channels, however, only relatively few simulation models have been developed, e.g. SHESED (WICKS & BATHURST 1996), ETC (MATHYS et al., 2003), SETRAC (RICKENMANN et al., 2006a), PROMAB (RINDERER et al., 2009), and sedFlow (HEIMANN et al., 2015a). These models are similar to numerical sediment transport models that were applied primarily in flatter mountain rivers, e.g. MORMO (SCHILLING & HUNZIKER 1995) and BASEMENT (VETSCH et al., 2005). However, experience with the use of such simulation models for the steeper channels (and especially for torrents) has been very limited. Especially regarding the two-dimensional simulation of bedload deposition on torrent fans, there is scarcely any experience, except for the recent study of CHIARI and SCHEIDL (2015).

The one-dimensional bedload transport model *SETRAC* (RICKENMANN et al., 2006a; CHIARI et al., 2010) was tested for its suitability in the case of steep channels by means of flume experiments (KAITNA et al., 2011) and well-documented bedload-transporting flood events in August 2005 in Switzerland and in Austria (CHIARI & RICKENMANN 2009, 2011). The simulation model was developed especially for use in torrent catchments and in mountain rivers, taking into account a reduced transport capacity due to high-energy losses caused by macro-roughness elements. *TomSed* is the follow-up model of *SETRAC*; it is freely available from http://www.bedload.at. To take into account increased energy losses for bedload transport a method for partitioning flow resistance was developed for *SETRAC*. This method is based on (only) 373 flow velocity measurements. It leads to a similar reduction of the energy slopes as with the other method (RICKENMANN & RECKING (2011) presented in chapter 2.4 for partitioning the flow resistance and based on 2890 flow velocity measurements. In a new version of *TomSed* the new approach of RICKENMANN & RECKING (2011) to partition flow resistance was also implemented. The one-dimensional bedload transport model *sedFlow* also includes the new approaches to calculate flow resistance and bedload transport in steep channels. Being based on rectangular cross-sections and with an option for using simplified hydraulics without flow routing, it requires only short calculation times. Thus the program *sedFlow* allows simulating many different scenarios or conducting sensitivity analyses with the variation of different input and model parameters in a relatively short time. The model *sedFlow* was calibrated with observations of bedload transport in several Swiss mountain rivers (HEIMANN et al., 2015b; RICKENMANN et al., 2014, 2015). The model *sedFlow* is freely available from http://www.wsl.ch/sedFlow.

Chapter 4

Debris flows

4.1 PROPERTIES OF DEBRIS FLOWS

In steep headwater catchments in the Alps debris flows occur every year, and frequently such events cause considerable damage. A debris flow is a rapidly flowing mixture of soil with different amounts of water. At the front of a debris flow there is a high concentration of solids, and these flows are characterized by an unsteady and surging flow behavior, clearly distinguishable from a typically more steady water discharge in a stream channel.

The grain composition of debris flows can vary considerably. In the Alps coarse blocks frequently collect at the front of a debris flow. Coarse particles are often also transported in the other parts of a debris-flow surge. In the case of channel overtopping on the fan, a lot of coarse debris may be deposited there. This type is also termed a *granular debris flow* (Fig. 4.1). In the case of mudflows (Fig. 4.2) the fine material and the water dominate, whereas generally the coarse stones and blocks are missing or they have a negligible influence on the flow behavior. In the rearward part of a debris flow (or mudflow) the solids concentrations are usually smaller than in the front part (Fig. 4.3). The deposition conditions are then similar to those caused by the processes of fluvial bedload transport outside of a channel (COSTA 1988; HÜBL et al., 2002; PIERSON 2005).

Depending on the material composition, different theoretical approaches were proposed to describe the flow behavior. However, a partitioning into various flow types based on field observations is often only possible in a rudimentary way. This difficulty of a simple identification of different flow types is also reflected in the terminology and the classification of debris flows, and in different languages there are some differences in meaning. A rough correspondence of the terms in German, French, Italian and English is presented in Table 4.1.

Hillslope debris flows (HUNGR et al., 2001, 2014; HÜRLIMANN et al., 2015) are distinguished from debris flows basically by the place where they occur (terrain conditions with weakly or no predetermined lateral limits to the flow path) and often by relatively short flow distances, while the latter flow type typically runs in a channel or a gully. Generally, hillslope debris flows do not occur several times at the same place and also do not exhibit multiple surges. In the early stages, hillslope debris flows (Fig. 4.4) can be compared with spontaneous shallow landslides, after a larger

Figure 4.1 Front of a granular debris flow, Kamikamihori valley, Japan (photo H. SUWA).

Figure 4.2 Front of a mudflow, Jiangjia valley, China (photo Z. WANG).

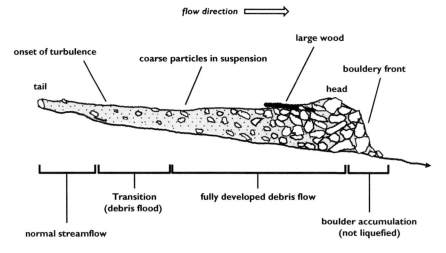

Figure 4.3 Typical longitudinal section through a debris flow with decreasing solids concentration from the front to the rearward part. Adapted from PIERSON (1986).

Table 4.1 Terms for debris flow in a few languages.

German	French	Italian	English
(Granularer) Murgang	Lave torrentielle	Colata detritica, lava torrenzia	Debris flow, granular or stony debris flows
Schlammstrom	Coulée de boue, lave torrentielle boueuse	Colata di fango	Mud flow
Hangmure	Coulée de boue de versant	Colata detritica di versante	Hillslope debris flow (debris avalanche)

Figure 4.4 Example of hillslope debris flows (Sachseln, Switzerland). These often occur in non-forested areas (photo: Oberforstamt Obwalden).

displacement distance the flow behavior may be similar to that of debris flows. In rainstorm events shallow landslides may transform into hillslope debris flows.

The typical debris flows in the Alps can be considered in a simplified way to be a mixture of the three main components water, fine material and coarse granular material. Based on their composition and flow behavior, debris flows are a mixture of floodwaters, landslides and rock slides or debris avalanches. Fig. 4.5 shows the relative proportions of the three main components for such rapid mass movements. Therefore, the physical processes in the formation, the flow and the deposition of debris flows are correspondingly complex and are only partially understood.

In comparison with floods with fluvial bedload transport in torrent channels, debris flows have greater flow depth, may cause greater erosion and entrainment of solid material, thus often transporting large amounts of debris to the fan or confluence area. During a flood, particles are moved along the channel by the driving force of the water. Debris flows with high solids concentrations typically exhibit a much greater viscosity or frictional resistance than just water alone. For the triggering of debris flows, a minimum amount of granular material is required in addition to water, as well as steep slopes. The most important properties of debris flow and of traces that are left behind on the terrain are summarized in Table 4.2.

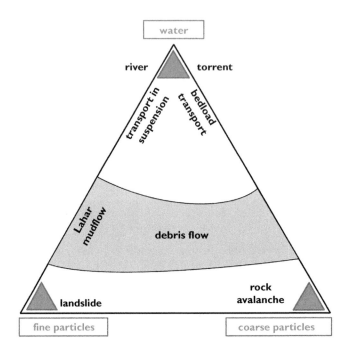

Figure 4.5 Main components of a debris flow in a three-phase diagram, in comparison with other rapid mass movements. Modified from Phillips & Davies (1991).

Table 4.2 Characteristic properties of debris flows (after Rickenmann 1996).

Material composition and flow behavior:
- wide range of grain sizes, +/− uniformly distributed over flow depth
- transport of very large blocks possible
- highest bedload concentration at the front, further back often a more fluid mixture
- high bulk density of the water-solid mixture (ca. 1.6 … 2.4 t/m³ at the front)
- high "viscosity" (non-Newtonian fluid)
- mostly discharge in waves, i.e. intermittent advance of one or more surges
- formation of debris walls (levées) in absence of already existing boundary of the flow cross-section, e.g. by steep rock wall
- deposition especially at locations with widening of stream bed or with sudden decrease of channel slope; in Alpine channels often at slopes of 5% … 18%

Characteristic traces in terrain, which point to debris flow activity:
- levées (lateral debris walls)
- residual debris tongues, on the fan or in the case of local channel widening
- unsorted deposition (all grain sizes well mixed, no layering)
- coarse blocks and fine-grained matrix (if not yet washed out) in the deposits
- polished and scarred areas of rock
- clear boundary of the deposits
- often little damage to vegetation in the deposition areas outside of the channel
- erosion or flow cross-section usually U-shaped

If the banks of a channel are shallow, then natural levées tend to form due to material deposition along the path of the debris flow, in a manner that createstheir own lateral boundary of the flow cross-section. The debris is deposited typically in flatter terrain and in a non-uniform way. During the depositional phase, the high viscosity or the large grain-to-grain friction of the decelerating mixture leads to a relatively abrupt stop of the frontal part of the debris flow. Thus, the front part of the deposits is usually demarcated clearly from the old terrain. The irregular deposition of multiple surges results in a rough terrain surface of a debris-flow fan. Observers of debris flows report that such events are often accompanied by a loud noise, ground vibrations and sometimes also by a sulfuric smell. Occasionally these phenomena have also already been noticed shortly before the arrival of the debris flow. A more detailed description of the debris flow process may be found, e.g., in Costa (1984, 1988).

Typical debris-flow parameters are summarized in Table 4.3, as estimated for the biggest events during two major rainstorms of summer 1987 in Switzerland. Table 4.4 gives an overview of characteristic properties of debris flows compared with floods with sediment transport and with debris flow-like discharges in the transition zone. The debris load is usually estimated from the volume of the deposits for a whole event and, thus, also contains the pore volume.

The maximum discharge usually occurs near the debris flow front. With alpine debris flows, the maximum discharge may be 100 m³/s to 1000 m³/s, and, therefore, is about 10 to 100 times greater than a comparable peak flood discharge in the same torrent channel (Table 4.3). The height of the debris flow front can amount to 10 m, and flow velocities up to 15 m/s (54 km/h) have been estimated for alpine debris flows. With larger debris flow events in the Alps, a few 10,000 m³ to several 100,000 m³ of sediment could be deposited on the fan. A substantial amount of material is eroded sometimes in the area of the fan. The total runout distance depends, among other things, on the amount of material transported for each surge.

Table 4.3 Typical debris-flow parameters of the biggest events in the summer of 1987 in Switzerland (after Zimmermann & Rickenmann 1992). DQ = data quality: **** = very good, *** = reliable, ** = rough estimate, * = very rough estimate/uncertain traces.

Debris flow event, Date:	Val Varuna 18.7.87 Value	DQ	Val da Plaunca 18.7.87 Value	DQ	Val Zavragia 18.7.87 Value	DQ	Minstigertal 24.8.87 Value	DQ
Debris load [m³]	200'000	****	250'000	****	30'000	**	30'000	***
Maximum flow velocity at the fan apex [m/s]	8	**	10	*	8	***	14	**
Flow depth at the fan apex [m]	6	***	?		6	****	10	***
Maximum discharge [m³/s]	400–800	**	400–900	*	500–700	****	150–250	***
Peak discharge of water only, estimated [m³/s]	7	*	9	*	30	**	17	**
Number of surges	ca. 10	**	>5	*	< = 6	**	1	***
Max. load per surge [m³]	50'000	**	80'000	**	<30'000	**	<30'000	***
Max. erosion depth [m]	11	****	12	****	2	**	4	**
Max. erosion cross-section [m²]	650	****	550	****	20	**	55	**
Historical events	ca.10 in 150 years	****	none known	***	ca. 7 in 150 years	***	uncertain, unknown	*

Table 4.4 Overview of the properties of characteristic displacement processes in torrents (after Pierson & Costa 1987; Costa 1988; Hungr et al., 2001; Pierson 2005; Hübl et al., 2006).

Process type	Flood		Debris flow	
Terms (German)	Hochwasser	Fluvialer Feststofftransport	Murgangartiger Feststofftransport	Murgang
Terms (English)	Flood	Bedload transport	Debris flood (hyper-concentrated flow, immature debris flow)	Debris flow
Process type	Discharge of water only	Weak bedload transport	Strong bedload transport	Debris flow
Flow behavior	Newtonian	Newtonian	Approx. Newtonian	Non-Newtonian
Vol. solids concentration (approx. range)	Per mill range	0–20%	20–40%	>40%
Max. grain size	mm–cm	dm	m	M
Density (approx. range)	1000 kg/m³	1300 kg/m³	<1300–1700 kg/m³	>1700 Kg/m³
Viscosity (approx.)	0.001–0.01 Pas	0.01–0.2 Pas	0.2–2 Pas	>2 Pas
Shear strength	None	None	None	Present
Relevant acting forces	Turbulence, bed shear stress	Turbulence, bed shear stress	Buoyancy, turbulence, bed shear stress, dispersive pressure	Buoyancy, dispersive pressure, viscous and frictional forces
Vertical distribution of solid particles over flow depth		Coarser particles near the bed (rolling, hopping, jumping) and suspended sediment distributed in cross-section	Solids and suspended sediment distributed in cross-section	Solids distributed in cross-section
Deposition characteristics		Horizontal or inclined stratification; coarser clasts may be imbricated	Weak horizontal stratification; mostly grain supported	Terminal debris lobes; marginal levées and tongue-shaped deposits; grain or matrix supported; usually clear boundary of deposits, U-shaped channels
Sorting of the deposited solids	(yes)	Sorting moderate to good within individual bedding units	Moderate to poor sorting	Non-stratified, extremely poorly sorted

4.2   IMPORTANT ELEMENTS OF THE PROCESS AND HAZARD ASSESSMENT

Standardized procedures and regulations for the management of natural hazards were introduced, for example, in Europe during the last two decades (e.g. HÜBL et al., 2002; PETRASCHECK & KIENHOLZ 2003; GREMINGER 2003; FUCHS et al., 2008; HÜRLIMANN et al., 2008). These regulations require the determination of hazard danger levels for potentially affected areas such as a fan. The hazard danger levels are a function of process intensity and probability of occurrence. For debris flows and floods, process intensities are typically defined as a function of flow velocity and flow depth, both of which vary spatially and depend on the magnitude or peak discharge of the process (HÜRLIMANN et al., 2008). The sediment volume or the sediment-water volume of a debris-flow event or of a single surge is typically taken as a measure of the magnitude (JAKOB 2005).

Thus, for the process and hazard assessment of debris flows—similar to other gravitational natural hazards—two key aspects need to be investigated: (a) the probability of occurrence (or return period) and the magnitude of the event (magnitude-frequency relationship), and (b) the flow and deposition behavior. The most important elements and existing dependencies are presented schematically in Fig. 4.6. The topic "magnitude-frequency" of torrent events is considered in chapter 5. Other important aspects are discussed below, where a range of methods and approaches are presented. A further section is devoted to a brief overview of GIS-based and numerical simulation models. At the end of chapter 4 the depositional behavior on the fan is discussed, which is often very important for the hazard assessment.

The most important questions in relation to the hazard assessment are:

- What *event magnitude* has to be expected?
- What *probability of occurrence* has to be considered?
- Which are the *endangered areas*?

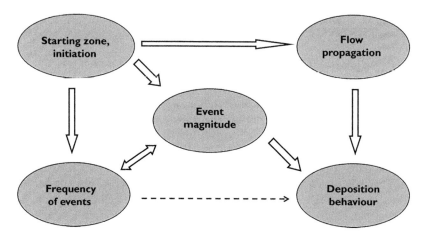

Figure 4.6 Most important elements in the assessment of debris flow events, and the dependencies between the elements (after RICKENMANN 2001b).

In a somewhat simplified way, the two key aspects can be grouped as follows:

a) Magnitude and Frequency
   Here, the following three primary elements have to be considered:
   • Initiation (location, type of triggering mechanism)
   • Event magnitude
   • Event frequency

b) Flow and depositional behavior
   Here, the following three primary elements have to be considered:
   • Event magnitude
   • Flow behavior in the channel
   • Depositional behavior on the fan

As may be seen from a detailed process assessment of debris flow events, the above subdivision into two key aspects represents a simplification. The debris flow can entrain and accumulate additional material during the flow process (from the stream bed, the banks, and the side slopes) or also deposit the material again. With the present state of knowledge it is difficult to quantify these processes reliably. Thus, in an initial step an event magnitude is frequently estimated for the location of the fan apex, and this value becomes an important input quantity in the assessment of the flow behavior further downstream. This simplification may be a reasonable approximation if the debris-flow parameters are primarily needed for estimating potential hazards in the area of the fan where, in many cases, no significant material entrainment takes place.

For the hazard assessment many different methods and approaches are available. This variety reflects, on the one hand, the different characteristics of various debris flow types and, on the other, the limited state of knowledge. A tabular overview of the available methods to determine the important elements is given in RICKENMANN (2001b, 2015).

4.3   OCCURRENCE OF DEBRIS FLOWS

4.3.1   Predisposition for debris flow occurrence

The assessment of the possibility of a debris flow occurring in a torrent should be based primarily on an interpretation of the fan area as well as on the traces of earlier events and/or historical information. If these produce no clues, some general characteristics of the catchment area may be used to allow a rough assessment (Table 4.5). A minimum streambed slope and a sufficiently large bedload potential are the necessary requirements for debris flows to occur at all. These two factors are also the most important criteria to assess the hazard potential of a torrent. For the formation of a debris flow from the channel bed or from a hillslope, the minimum slope is approximately 25–30%. With the presence of other factors promoting the development of a debris flow (e.g. channel constrictions, driftwood), a debris flow can also form in the case of slopes of approximately 15–25%. With channel slopes less than 15% the development of debris flows is not likely.

Table 4.5 Influence of channel slope (*S*) and bedload potential (*F*) on debris-flow hazard. Significance of hazard classes: A1: high debris-flow hazard, A2: medium debris-flow hazard, B: low debris-flow hazard, C: practically no debris-flow hazard (from RICKENMANN 1995).

Triggering zone: slope of stream bed or hill-slope	Channel features and bedload potential F (channel + hillslopes)	Hazard class
$S > 25\%$	Channel in granular material, potentially larger slope instabilities ($F > 10\ 000$ m³)	A1
	Channel mainly in granular material ($F = 1\ 000 - 10\ 000$ m³)	A2
	Channel mainly in bedrock ($F < 1\ 000$ m³)	B
$15\% < S < 25\%$	Channel in schistose, flysch-like rocks, potential slope instabilities ($F > 10\ 000$ m³)	A1
	Other rock types, channel with possible log jams ($F > 10\ 000$ m³)	A2
	Channel without possible log jams ($F = 1\ 000 - 10\ 000$ m³)	B
	Channel mainly in bedrock ($F < 1\ 000$ m³)	C
$S < 15\%$	Not relevant	C

The classification of the debris flow potential in Table 4.5 is based on the analysis of the debris flows in the Swiss Alps in 1987 (RICKENMANN & ZIMMERMANN 1993) as well as on a semi-quantitative assessment of the debris flow hazard after AULITZKY (1973) and NAKAMURA (1980). The hazard classes in Table 4.5 correspond to a mix of a probable intensity of the events and a possible frequency of occurrence. The statement, however, refers primarily to the expected maximum intensity of an event. Due to the large bedload potential and the steep slopes in hazard class A, for example, small debris flows could also occur. This outcomeleads to a higher overall frequency than, for instance, for class C, where only small debris flows are to be expected. (The significance of the hazard classes is given in the legend in Table 4.5.)

A rough differentiation between the processes of debris flow and bedload transport can be made, based on the morphometric parameters of the catchment and of the fan. Therein, the mean channel slopes on the torrent fan, S_f, are plotted as a function of the MELTON number, *Me*, defined as the difference between the highest and the lowest elevation values, normalized with the square root of the catchment area (MARCHI & BROCHOT 2000; BARDOU 2002; RICKENMANN & SCHEIDL 2010). Larger values for S_f and *Me* define the range of occurrence of debris flows, while smaller values define the occurrence of bedload transport. However, the demarcation between the two ranges is not very clear, but there is quite a wide transition range (Fig. 4.7). Other classification schemes were proposed for example in terms of catchment length and Melton number (WILFORD et al., 2004).

4.3.2   Triggering conditions

The triggering of debris flows can take place in the form of hillslope instabilities or of channel destabilization. With more intensive rainfall there is an increased probability for shallow landslides to occur in steep terrain. There are quite a large number of studies which focused on (critical) rainfall conditions necessary for the occurrence

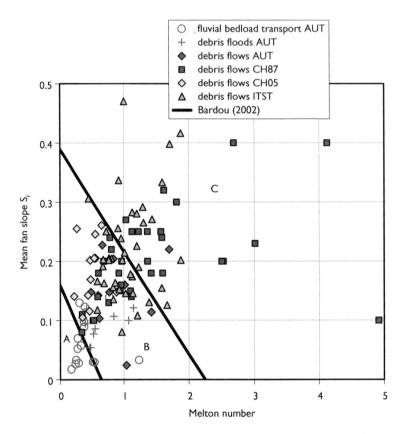

Figure 4.7 Rough demarcation of torrents capable of debris flow and those with fluvial bedload transport (modified from Scheidl & Rickenmann 2010), based on the mean fan slope S_f and the Melton number Me. The data come from Switzerland (CH), Austria (AUT) and South Tirol in Italy (ITST). The zones A (fluvial transport), B (transition range) and C (debris flow) correspond to the classification of Bardou (2002), who also used data from Marchi & Brochot (2000).

of shallow landslides or hillslope debris flows. Regarding debris-flow initiation from massive channel erosion, however, there are only relatively few studies that attempted to define a limiting discharge (analogous to the start of bedload transport) based on laboratory flume investigations as well as on simple theoretical estimates. These methods, however, usually provide only a rough estimate of the triggering rainfall conditions or the limiting discharge for the formation of the debris flow, since the influence of the properties of the hillslope material or of the streambed material is generally not taken into account.

In the case of more intensive and prolonged snow melt, the tendency for debris-flow formation is increased with the increasing ground saturation. For bigger debris-flow events in the alpine regions, however, more intensive rainfall is often needed in addition. Water plays an important part in the initiation process. In the high alpine scree the destabilization can already be brought about by an underground saturation of the granular material. Since the surface discharge is not the only factor for the triggering process, not only is the rainfall intensity of importance, but also the extent of ground saturation due to prolonged rainfall. In the case of storm rainfall in Switzerland, for example, a minimum triggering intensity of around 30 mm/h and a minimum total rainfall of around 40 mm have to be reached at the same time for the formation of debris flows (ZIMMERMANN et al., 1997). In the inner-alpine regions of Switzerland somewhat smaller triggering rainfalls are necessary for debris-flow formation; this requirementcould be connected with the smaller annual rainfall compared with the areas bordering the Alps (ZIMMERMANN et al., 1997). Estimates for critical rainfall conditions are frequently expressed in terms of mean rainfall intensity I [mm/h] and the duration D_R [h] of the triggering rainfall event. These limiting conditions can vary strongly both regionally and locally. Critical rainfall conditions for debris-flow occurrence in Switzerland are shown in Fig. 4.8 with data from Austria in comparison with threshold lines for related processes in regions close-by. Threshold rainfall conditions for slope instabilities have been determined for many regions in the world (e.g., CAINE 1980, GUZZETTI et al., 2007, 2008).

Debris flows can occur in torrent channels if sufficient solids material is available in the bed. For the formation of debris flow, a minimum amount of granular material needs to start moving with a relatively high solids concentration. This process occurs primarily in steep channels and in constricted places with previous obstruction of the material flow, possibly with a temporary clogging of the flow cross-section, or with an abrupt increase of the erosion in the channel. Fig. 4.9 shows the difference between the process of debris flow and other types of solids displacement in and near steep channels and its connection with the formation of debris flows.

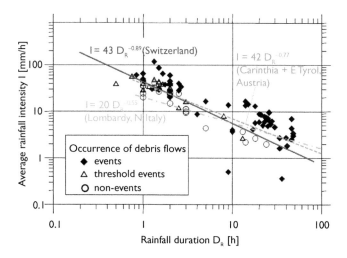

Figure 4.8 Empirical relationships for critical rainfall conditions for the triggering of debris flows and landslides. The data for debris-flow occurrence in Switzerland as well as the corresponding limiting criterion are taken from ZIMMERMANN et al. (1997). The threshold lines for southern and eastern Austria (soil slips) as well as for northern Italy (landslides) are taken from a compilation of GUZZETTI et al. (2007). The three geographic regions are all nearby.

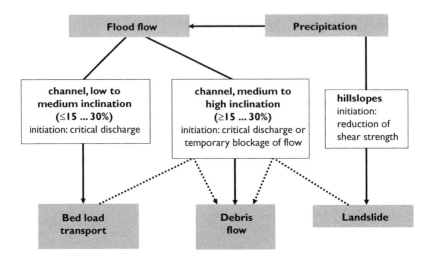

Figure 4.9 Significant bedload delivery processes in torrents and the role played in the formation of debris flows. Modified from RICKENMANN (1996).

A simple analysis of slope stability leads to a theoretical limiting slope angle of about 12° to 17° for typical conditions in the stream bed, assuming a friction angle of the material in the range 33° to 37° (TAKAHASHI 1987). In steep headwater channels (as also in rivers) a minimum discharge is necessary for bedload to be transported (see critical discharge in Fig. 4.9). The combined loading of the bed due to the discharge and the bedload in motion can suffice in steep slopes to cause enough solids to move suddenly due to a channel-bed destabilization. The resulting debris-flow formation by channel erosion appears to depend both on the slope at the initiation area and the discharge, as indicated by the data in Fig. 4.10. An example of the formation of a debris flow from the channel is shown in Fig. 4.11.

If shallow landslides are sufficiently large and fluid or occur very near to channels, they may transform into hillslope debris flows, get into the torrent channel and continue moving downstream as a debris flow. Shallow landslides that occurred in Switzerland in the period 1997 to 2005, exhibit typical slope angles from about 24° to 43° and volumes of 50 to 100 m³ (RAETZO & RICKLI 2007). The rainfall conditions for triggering shallow landslides in Switzerland were investigated by RICKLI et al. (2008).

A continuous transition from fluvial bedload transport to debris floods to debris flows may be expected where channel slopes become steeper than about 20–25%, according to investigations of SMART & JÄGGI (1983) and RICKENMANN (1990) and other flume and field observations. A similar conclusion may be drawn by comparing critical discharge criteria for bedload-transport initiation and debris-flow formation. Analogous to the limiting discharge at the start of fluvial bedload transport, a critical dimensionless discharge q_c^* for the formation of debris flows can be defined:

$$q_c^* = q_c/[g^{0.5} \, D^{1.5}] = a_g/S^{\alpha g} \tag{4.1}$$

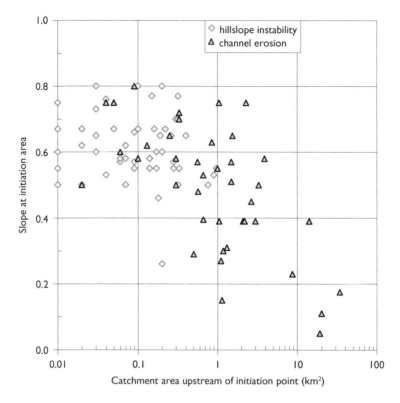

Figure 4.10 Hillside and channel slopes in the initiation area of debris flows that formed either due to hillslope instability or channel erosion (channel destabilization) are shown as a function of the catchment area above the triggering zone (as indicator for the water discharge). Data taken from Swiss investigations (VAW 1992; ZIMMERMANN et al., 1997).

Figure 4.11 (a) Channel above the triggering zone (corresponds to the situation in the figure on the right before the initiation of debris flow); (b) Channel at the triggering zone after the initiation of a debris flow. The channel slope is 51% (S = 0.51). (Photos M. ZIMMERMANN).

Here q_c is the critical specific discharge per meter channel width, D is a characteristic grain size in the channel bed, $S = \sin\beta$ represents the channel slope, g = gravitational acceleration, a_g = empirical coefficient, and α_g = semi-empirical exponent.

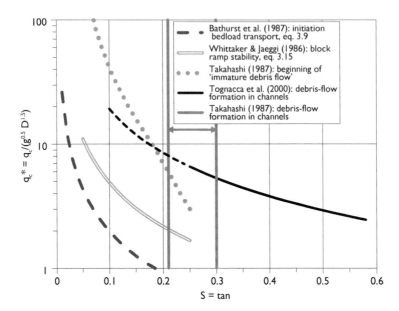

Figure 4.12 Relations based on flume experiments to determine the critical unit discharge for the initiation of different process types. The criteria for the formation of debris flow by channel erosion have not yet been checked against field data.

Such relations to determine the critical unit discharge for the initiation of different process types are shown in Fig. 4.12. The relations of TOGNACCA et al. (2000), WHITTAKER & JAEGGI (1986), and BATHURST et al. (1987) are all based on laboratory flume tests. Fig. 4.12 also show a limiting condition for the start of "immature debris flows" after TAKAHASHI (1987), a state which corresponds roughly to intensive bedload transport or "debris flood" conditions; this relation is also based on flume experiments. The figure shows a large range of possible limiting discharges for the formation of debris flows in channels that still have to be checked against field observations.

4.4   EMPIRICAL APPROACHES TO CHARACTERIZE THE FLOW AND DEPOSITION BEHAVIOR

To assess the flow and deposition behavior either empirical approaches including estimate formulae or numerical simulation models can be used. The main objectives are to determine critical locations where there is the possibility of flow overtopping a channel and to delineate areas of the fan that are likely to be inundated and covered by solid deposits. A summary of the debris flow parameters estimated for events in 1987 in Switzerland is given in Table 4.3. More exact measurements have been made for several years at several debris flow monitoring stations in the Alps (e.g. GENEVOIS et al., 2000; MARCHI et al., 2002; RICKENMANN et al., 2001; HÜRLIMANN et al., 2003; MCARDELL et al., 2007).

The sequence of some simple empirical calculations using important parameters for characterizing debris-flow behavior is illustrated schematically in Fig. 4.13. If, for a future event, the debris load M is estimated, an approximate maximum discharge

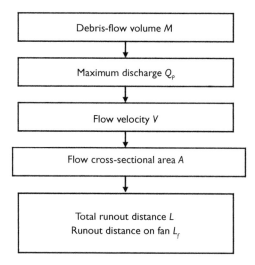

Figure 4.13 Calculation sequence of empirical approaches to estimate the most important flow parameters of debris flows. Modified from RICKENMANN (1999).

Q_p can be then be calculated. This value together with the channel slope determines essentially the flow velocity V. The maximum required flow cross-section A is then given by $A = Q_p/V$. A comparison with the existing channel cross-section gives indications of possible places of channel overtopping. Also, the total runout distance of a debris flow from the point of initiation to the lowest deposition point, L, or the deposition length on the fan, L_f, can be estimated roughly based on the debris load, if a more exact determination by means of simulation models is not possible.

The observed values of debris load usually contain both the bedload and also the pore or water volumes. If the data were obtained from a measuring station, then M is typically determined by the integration of the mixed discharge over time. In other cases M is determined from the observed deposition area and the (mean) deposition thickness, whereby the pore volume is included, which corresponds approximately to the water content.

Regarding the erosion behavior along the transit stretch, relatively few quantitative observations were made in the field (HUNGR et al., 2005; SCHÜRCH et al., 2011; BERGER et al., 2011a; McCoy et al., 2012). Besides using modelling concepts based on soil mechanics (IVERSON 2012; McCoy et al., 2012), analogies were made with approaches developed for bedload transport (EGASHIRA et al., 2001; RICKENMANN et al., 2003; CAO et al., 2004). In estimating the debris load using geomorphologic methods, therefore, possible solids input from the transit stretch are included, if the estimation procedure is applied in the fan area according to Fig. 4.13.

4.4.1 Maximum discharge

Field observations indicate an empirical relation between maximum discharge Q_p [m³/s] and debris load M [m³] (RICKENMANN 1999), as illustrated in Fig. 4.14. Here one distinguishes between granular and muddy debris flows (MIZUYAMA et al., 1992):

$$Q_p = 0.135 \ M^{0.78} \hspace{3cm} \text{(granular debris flow)} \quad (4.2)$$
$$Q_p = 0.0188 \ M^{0.79} \hspace{3cm} \text{(muddy debris flow)} \quad (4.3)$$

This distinction is based on observations made in Japan, and it has been confirmed partly by debris flow data from other parts of the world. The classification, however, is not always easy. Thus, the data of the Rio Moscardo in Fig. 4.14 possibly lies nearer to the line of Eq. 4.3, since the debris flow discharges may have been rich in water. The debris flows in the Jiangjia ravine are very rich in fine material and one would expect therefore that the data points would scatter around Eq. 4.3 rather than Eq. 4.2. The debris flows in the Illgraben are also rich in fine material, but contain, in general, little cohesive sediment. When using the equations for predictions, the data range of the observed values should be considered as well as the fact that the maximum discharge should be correlated preferably with the volume of the biggest single surge than with the total debris load of an event. In the case of the data of the debris flows "Switzerland 1987" plotted in Fig. 4.14, the individual surge volume is not known. In alpine regions the assumption that M is greater than about 50,000 m³

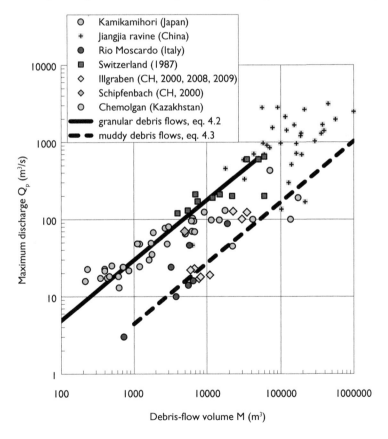

Figure 4.14 Maximum discharge Q_p of the water-solids mixture as a function of the debris load M. Data sources are indicated in RICKENMANN (1999), and some more recent data are from MARCHI et al. (2002), HÜRLIMANN et al. (2003), RICKENMANN et al. (2003), and BERGER et al. (2011b).

is rather implausible, as shown in the analysis of the debris-flow surges of the 1987 events in Val Varuna (VAW 1992).

4.4.2   Flow velocity

To estimate the mean flow velocity V [m/s] two different equations are proposed here (*RICKENMANN 1999*):

$$V = 2.1 \, Q^{0.33} \, S^{0.33} \tag{4.4}$$
$$V = k_{St} \, h^{0.67} \, S^{0.5} \tag{4.5}$$

where Q [m³/s] is the discharge, S the channel slope [m/m] in the considered stream reach, k_{St} a pseudo STRICKLER coefficient [m$^{1/3}$/s] and h [m] the flow depth. The use of Eq. 4.4 and Eq. 4.5 for data from debris flows and water discharges are shown in Fig. 4.15 and Fig. 4.16. In these figures only some datasets from RICKENMANN (1999) were used, i.e. datasets A and B (debris-flow data with directly measured flow velocities) and dataset G.

For the debris-flow data shown in Fig. 4.16, an average friction value is about $k_{St} = 10$ m$^{1/3}$/s. For granular debris flows in natural channel reaches, k_{St} values are obtained in the range of 6 m$^{1/3}$/s (RICKENMANN & WEBER 2000). For debris flow discharges in artificial (canalized) channels, the pseudo MANNING-STRICKLER coefficients could be up to 50% higher (PWRI 1988). Eq. 4.4 is valid for natural channel reaches; thus, in artificial channels an approach such as Eq. 4.5 is preferable.

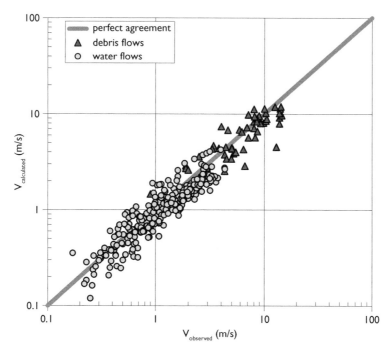

Figure 4.15 Use of Eq. 4.4 for debris flows and water discharges. Comparison of calculated and observed flow velocities. Modified from RICKENMANN (1999).

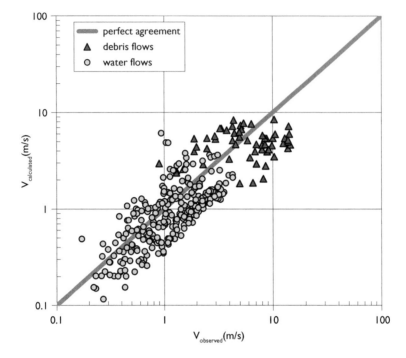

Figure 4.16 Use of Eq. 4.5 for debris flows and water discharges. For both sets of data in each case, a mean STRICKLER coefficient k_{St} is assumed. Comparison of calculated and observed flow velocities. Modified from RICKENMANN (1999).

4.4.3 Total runout distance

If there are significant deposition or redistribution reaches, a small debris flow could come to a standstill there. The estimate of the possible deposition volumes in such a reach and the comparison with the estimated expected debris load permits an assessment of whether discontinuation of the debris flow is probable. Depositions and overflowing of the banks may also take place upstream of narrow flow cross-sections as a result of retrogressive aggradation. In the case of large debris-flow events (which are relevant for the hazard assessment), it is generally very likely that debris flows reach to the fan. The analysis of 82 debris-flow events in the summer of 1987 in Switzerland showed that: (a) A minimum general slope of less than 19% was nowhere found. The general slope denotes the mean slope of the total flow path from the point of initiation of the debris flow down to the lowest deposition point; (b) In general, the runout distance L [m] is dependent on the debris load M [m³]. The analysis of further data from other regions suggested the inclusion, in addition, of the height difference H_e [m] between the uppermost point of initiation and the lowest deposition point (RICKENMANN 1999; RICKENMANN & SCHEIDL 2010), resulting in the following equation for the mean runout distance:

$$L = 1.9 \, M^{0.16} \, H_e^{0.83} \tag{4.6}$$

To estimate an upper limit of the runout distance L_{max} the following relation may be used:

$$L_{max} = 5 \; M^{0.16} \; H_e^{0.83} \tag{4.7}$$

In contrast to Eq. 4.2 and Eq. 4.3 for the derivation of Eq. 4.6 and Eq. 4.7 the total debris load was used and, thus, it has to be input here. The use of Eq. 4.6 and Eq. 4.7 for data from debris flows is shown in Fig. 4.17. For predictive estimates, a further relation between L and H_e is necessary. This is the longitudinal profile of the expected flow path, whereby Eq. 4.6 and Eq. 4.7 can be solved either mathematically or graphically.

Other empirical equations for the total travel distance of debris flows were proposed by COROMINAS (1996), LEGROS (2002), TOYOS et al. (2008), and PROCHASKA et al. (2008). In most of these approaches the runout length is essentially a function of

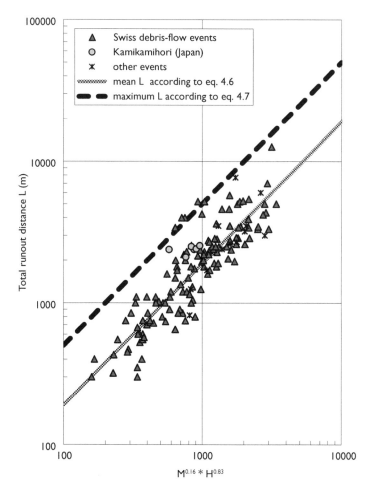

Figure 4.17 Estimation of the total runout distance with Eq. 4.6 and Eq. 4.7. The field debris-flow data are from RICKENMANN (1999). Modified from RICKENMANN (1999).

the volume and angle of reach or the longitudinal profile of the expected flow path. Another empirical approach to estimate the total travel distance is based on a sediment budget along the flow path (CANNON 1993; FANNIN & WISE 2001).

4.4.4 Deposition length of the fan

For the deposition length L_f on the fan the data used by RICKENMANN (1999) shows only a very weak dependence on the debris load M. Thus, an empirical estimation formula is not recommended. Nevertheless, for a particular torrent channel and a given fan topography, it may be expected that with similar material properties larger debris flows flow further than smaller ones. Larger debris flows have a tendency for larger maximum discharges (see Eq. 4.2 and Eq. 4.3), together with larger flow velocities and/or larger flow cross-sections. Based on a momentum consideration of the flow of the water-debris mixture on a uniformly sloping surface, the deposition length L_f can theoretically be estimated as follows (HUNGR et al., 1984):

$$L_f = A_V^2/G \tag{4.8}$$
$$A_V = V_u \cos(\beta_u - \beta) [1 + (g h_u \cos\beta_u)/(2 V_u^2)] \tag{4.9}$$
$$G = g (S_R \cos\beta_u - \sin\beta) \tag{4.10}$$

where β = slope of the deposition reach, β_u = slope of the steeper inflow channel, V_u = flow velocity in the inflow channel, h_u = flow depth in the inflow channel, S_R = friction slope (sliding friction only), assumed to be constant in the runout reach, and g = gravitational acceleration. HUNGR et al. (1984) assumed that $S_R = 0.176 = \tan(10°)$ and obtained thereby good agreement between observed values of L_f for five debris flows in Western Canada and values calculated using Eq. 4.8. On the other hand, the use of Eq. 4.8 for debris flows in the Kamikamihori valley in Japan (with measured discharge parameters) resulted in better estimated values for L_f, provided $S_R \approx 1.1 \tan\beta$ is chosen instead of $S_R = \tan(10°)$. This holds likewise for the application of Eq. 4.8 to some debris flows in Switzerland in 1987, if flow depths obtained from field observations are determined and the flow velocities are calculated with a CHEZY equation or with Eq. 4.2 and Eq. 4.4 (RICKENMANN 2005b). Fig. 4.18 presents a comparison of calculated and observed deposition lengths.

It is interesting to note that the values calculated with Eq. 4.8 to Eq. 4.10 assume a minimum for flow velocities V_u between about 2 m/s and 4 m/s (RICKENMANN 2005b). A possible dependence of S_R on β is not surprising since the friction slope depends on the material properties of the debris flow, which is reflected roughly also in the fan slope. Other methods to estimate the runout distance of debris flows are discussed in RICKENMANN (2005b) and RICKENMANN & SCHEIDL (2010).

4.4.5 Impact forces

Several studies suggested that the impact pressure of debris flows on obstructions may be estimated in a similar way to that for the dynamic pressure for Newtonian water flows. In the region of the front of the debris flow, stones and large blocks of up to several meters diameter can be transported. In this case, locally higher pressures are to be expected. Based on measurements of the debris flow impact force on rigid obstructions, it is estimated that the pressure is, on average, about a factor 2 to 4

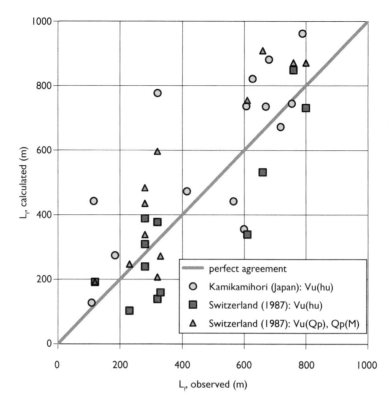

Figure 4.18 Comparison between calculated (Eq. 4.8 to Eq. 4.10) and observed deposition lengths
L_f using S_R = 1.1 tanβ. With the events in Switzerland in 1987 the flow velocity (V_u) was
estimated two ways: firstly as a function of the flow depth estimated in situ (h_u) with a
CHEZY equation for V_u (after RICKENMANN & WEBER 2000), and secondly by means of the
observed debris load (M) and Eq. 4.2 and Eq. 4.4. Adapted from RICKENMANN (2005b).

higher than the hydrostatic pressure (GEO 2000). Thus, the following formula for the
dynamic impact pressure p_d [N/m²] due to debris flow is proposed:

$$p_d = \alpha_d\, \rho_M\, V^2\, sin\beta_d \tag{4.11}$$

where ρ_M [kg/m³] = density of the debris flow mixture, V = flow velocity [m/s],
β_d = impact angle (often β_d = 90°) and α_d = empirical coefficient (for debris flow of
about α_d = 2 to α_d = 4). In flume tests the impact forces of both viscous and granular
debris flows were measured by SCHEIDL et al. (2013), and values for α_d ranging from
about 1.5 to 12 were determined. A recent experimental study on the impact force of
viscous debris flow indicated that the empirical coefficient α_d may be a function of the
approach flow FROUDE number Fr, with α_d = 5.3 Fr$^{1.5}$ (CUI et al., 2015). Interestingly,
in a flume study by SCHEIDL et al. (2015), a similar dependence on the FROUDE number was found for the correction coefficient in debris-flow velocity estimates using the
superelevation equation for Newtonian flows.

To estimate the impact force of granular debris flows, a theoretically-derived formula of Coussot (1997) resulted in a similar form to that for the dynamic pressure of Newtonian fluids, and the multiplication coefficient α_d may take on higher values than 2 to 4. To take into account the impact of individual large blocks on a structure, a computational method is described in Egli (2005). Further approaches for calculating the impact force of debris flows were proposed by Armanini & Scotton (1993), Zanuttigh & Lamberti (2006), and Ancey & Bain (2015). Rickenmann (2008) compared these methods for the range of flow depths and flow velocities that can be expected on the torrent fan with debris flows. It was shown that for Froude numbers *Fr* much smaller than 1 the hydrostatic component is important, whereas for *Fr* > 1 the hydrodynamic component dominates (see also Eq. 4.11).

4.5 MODELS FOR THE SIMULATION OF DEBRIS FLOWS

4.5.1 Empirical approaches

Some empirical methods or simple models to estimate the one-dimensional runout distance or deposition length on the fan are discussed in chapter 4.4.3 and 4.4.4. To delineate potentially endangered areas in more detail, the runout pattern or the surface area of potential debris-flow deposits on the cone should be known. A simple topography-based empirical approach was developed by Iverson et al. (1998), in which the deposition area of Lahars is correlated by means of an empirical scaling function with the event volume. With additional assumptions and the aid of a Geographic Information System (GIS), potentially endangered areas can be demarcated in a simple way. Similar methods were tested and implemented for debris flows by Hofmeister et al. (2003), Crosta & Agliardi (2003), Berti & Simoni (2007), Oramas Dorta et al. (2007), Griswold & Iverson (2008), and Scheidl & Rickenmann (2010). The key element of all approaches is an empirical scaling relation between the planimetric deposition area of debris flows and the event volume. As a modification, in the model TopRunDF much greater consideration is given to the fan topography using a random algorithm to determine a distribution of the potential flow paths (Scheidl & Rickenmann 2010). The model TopRunDF was tested using numerous debris flow events in Switzerland, in Austria and in South Tirol (Rickenmann & Scheidl 2010). TopRunDF can be downloaded from the web page: http://www.debris-flow.at.

4.5.2 Simple analytical methods

Zimmermann et al. (1997) and Gamma (2000) describe an automated application of simple models using a GIS technique. This method was used in several areas in Switzerland to produce hazard index maps. The zones of initiation of debris flows are determined basically from the limiting slope or limiting range of slopes. To estimate the flow velocity and the total runout distance, a mass point model for the flow behavior of a Voellmy fluid is used. The two friction parameters of the Voellmy approach must be estimated on the basis of earlier events (Rickenmann 2005b). The spreading on the fan is simulated by means of a stochastic algorithm, which also has to be calibrated beforehand. The bedload potential is estimated in a simplified way by accounting for potential sediment delivery areas. Depending on the expected debris load, a large num-

Figure 4.19 Comparison of the deposition of (a) a debris flow (event of 19.7.1987; Photo A. Godenzi, Chur) and (b) a snow avalanche (Photo R. Godenzi, Poschiavo, photo date 8.5.1978), which both occurred in the Val Varuna catchment in the neighborhood of Poschiavo (Canton Grisons, Switzerland). The two events have roughly comparable deposition volumes.

ber of simulation runs are calculated, and thereby the spreading over the fan is coupled implicitly with the event magnitude. The resulting model is called DFWalk. In Switzerland the parameter of the Voellmy model DFWalk was investigated with the help of the back-calculation of a total of 75 debris-flow events with volumes in the range of 3000 m³ to 450,000 m³ (ZIMMERMANN et al., 1997; GAMMA 2000; GENOLET 2002).

The Voellmy model was originally developed for the analysis of the flow behavior of snow avalanches (BARTELT et al., 1999), and is based partly on hydraulics methods. A similarity between the deposition of debris flows and of snow avalanches is illustrated in Fig. 4.19.

HUNGR et al. (1984) and TAKAHASHI (1991) present a simple analytical method to describe the flow distance of a constant debris flow stream in the outflow region on the fan. The method is based on the momentum equation and the assumption of constant friction losses along the runout zone (see Eq. 4.8 to Eq. 4.10); it was first developed for snow avalanches and then applied to debris flows. The main difficulty lies in the selection of an appropriate friction coefficient (RICKENMANN 2005b). This method was implemented in TopFlowDF, which otherwise exhibits similarities to TopRunDF (SCHEIDL & RICKENMANN 2011). In contrast to TopRunDF an empirical surface-volume relationship is not necessary, but instead with TopFlowDF the (empirically determined) friction coefficient is important.

4.5.3 Numerical simulation models

Interestingly, to describe the general kinematic flow characteristics of rapid gravitational mass movements such as snow avalanches, debris flows, rock avalanches and shallow landslides, to some extent similar modeling approaches were proposed.

This is particularly evident for the simpler dynamic approach determining the flow behavior of a single-phase bulk mixture represented by a mass-point model (e.g. SCHEIDL et al., 2013), discussed for debris-flow application in the previous chapter 4.5.2.

Kinematic flow parameters like flow velocities or dynamic impact forces are often needed for a more detailed hazard assessment. This typically requires the application of numerical simulation models, which represent a more physically-based description of the flow behavior of gravitational mass movements of solids-water mixtures. The kinematic flow characteristics of a debris flow depend, for example, on the topographical and surface friction conditions, the water content, the sediment size and sorting and the dynamic interaction between the solid and fluid phases of the debris-flow mixture (IVERSON 1997). Debris flows with high flow velocities often exhibit a fluid-like displacement behavior, whereas, during the initiation and deposition phases, soil mechanics aspects are more important.

To describe the material and flow behavior of debris flows, various approaches were proposed and implemented in numerical simulation models. An important element of many proposed models is an appropriate formulation for the constitutive behavior of debris flows. The main problem for practical hazard assessment is that there are no clear criteria as to which methods (or constitutive equations) can be best applied to the various debris-flow types encountered in nature.

Initially, one group of simulation models considered the debris-flow mixture to be a quasi-homogeneous fluid as a first approximation, enabling the flow behavior to be described by a rheological model. A rheological model provides a relation between the shear rate γ_s (= change in flow velocity/change in flow depth) and the applied shear stress τ. The laminar flow behavior of water is defined as Newtonian flow (see Fig. 4.20) and can be described with the formula:

$$\tau = \mu \gamma_s \tag{4.12}$$

where μ = the dynamic viscosity. The simplest model to describe the flow behavior of viscous debris flow is the so-called BINGHAM model (see Fig. 4.20):

$$\tau = \tau_B + \mu \gamma_s \tag{4.13}$$

The variable τ_B stands here for the shear strength—a second material parameter—which has to be overcome by the driving forces before a fluid deformation (flow) can occur. As is evident from Fig. 4.20, a series of further models were proposed to describe the flow behavior of debris flows. A "pseudo-plastic" rheology is shear thinning (i.e. the effective viscosity decreases with increasing shear stress); a "dilatant" rheology is shear thickening (i.e. the effective viscosity increases with increasing shear stress). A number of models are partly- or fully-based on a rheologic formulation for a Bingham or viscoplastic fluid (CHOI & GARCIA 1993; LAIGLE & COUSSOT 1997; FRACCAROLLO & PAPA 2000; IMRAN et al., 2001; MALET et al., 2004). The rheological properties of a real debris-flow mixture are difficult to determine. For the fine material of a debris flow, the rheological parameters were determined from laboratory measurements for some model applications (e.g. LAIGLE & COUSSOT 1997; IMRAN et al., 2001; MALET et al., 2004). However, it is much more challenging to deter-

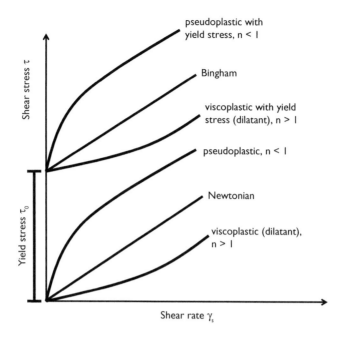

Figure 4.20 Rheological characterization of different fluids. *The* shear strength τ_B or τ_o is also termed yield stress in some studies. Modified from KAITNA (2006).

mine the effect of the coarser components on the rheology (PHILLIPS & DAVIES 1991; COUSSOT et al., 1998).

In several applications to natural debris flows, the pure Bingham model was modified by adding a friction term accounting for channel roughness and turbulence (O'BRIEN et al., 1993; HAN and WANG 1996; JIN and FREAD 1999). The model FLO-2D (O'BRIEN et al., 1993) has probably been the most widely applied, commercially-available, two-dimensional simulation program for debris flows. The constitutive equations consist of a rheological model that combines the BINGHAM rheology with an inertial friction term after BAGNOLD/TAKAHASHI (1991) as well as a turbulent friction term; the effects of the last two friction terms are lumped into an empirically determined pseudo MANNING coefficient (O'BRIEN et al., 1993). An example of the application of FLO-2D to simulate the deposition area on the fan of a debris-flow event in Switzerland is shown in Fig. 4.21, which also illustrates the effect of buildings, which can be considered optionally with this model. As a somewhat more complex alternative for a viscoplastic fluid, a HERSCHEL-BULKLEY model was implemented in another simulation program for debris flows (LAIGLE & COUSSOT 1996; RICKENMANN et al., 2006b).

With the second group of simulation models, the mass continuity for the water and the solids is considered separately, i.e. *two-phase models* are considered. The erosion and deposition of solids are taken into account using simple approaches. Such models were developed especially in Japan (e.g. NAKAGAWA et al., 2000). With

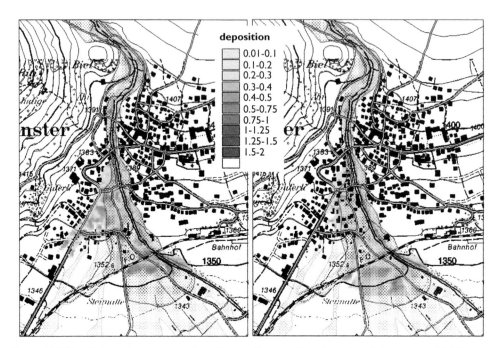

Figure 4.21 Simulation of the area and thickness of debris-flow deposits of the event of 24 August 1987 on the fan of the Minstiger stream (Switzerland) with the program FLO-2D. The area with dark-red points corresponds to the observed debris-flow depositions. In the figure on the right the effect of houses on the flow were taken into account in the simulation, but not in the figure on the left. In both cases the calculation was carried out with the same pseudo MANNING-STRICKLER value, but with different BINGHAM-parameters. The area with yellowish-green points in the lower fan region indicates fluvially redistributed finer sediment due to subsequent flooding.

the two-phase models, a discharge hydrograph can be used as input so that the resulting solids concentration depends basically on the channel slopes and the properties of the bed material. The deposition of the solids is obtained using similar methods as applied to fluvial bedload transport.

The modeling approach of IVERSON & DENLINGER (2001) takes account of basal pore water pressures and other soil mechanics aspects. The two phases of granular solids and a viscous fluid are coupled using mixture theory (IVERSON & DENLINGER 2001; DENLINGER & IVERSON 2001). The model is based on a generalization of the approach of SAVAGE and HUTTER (1989) for dry granular avalanches. A further development is the D-Claw model (IVERSON & GEORGE 2014; GEORGE & IVERSON 2014), which combines continuum conservation laws with concepts from soil mechanics, fluid mechanics, and grain–fluid mixture mechanics. An important aspect of this model is that both the solid volume fraction and basal pore-fluid pressure can evolve over time. It has been successfully applied to the 2014 landslide/debris flow event near Oso, Washington, USA (IVERSON et al., 2015).

The DAN model was derived from the work of Hungr (1995) for the analysis of the one-dimensional flow behavior of mass movements, with an option to select different rheological "friction" approaches. Similar simulation models were developed by Rickenmann & Koch (1997) and Näf et al. (2006). The DAN model was later extended to two-dimensional analyses and is designed to be an efficient tool for practical application (McDougall & Hungr 2004; Hungr & McDougall 2009).

The Voellmy approach is well known in Switzerland, above all due to its application to snow avalanches. It involves a base (Coulomb) parameter and a "turbulent" friction parameter (Bartelt et al., 1999). Numerical models with Voellmy rheology were successful in back-calculating shallow landslides, hillslope debris flows and channelized debris flows (Rickenmann & Koch 1997; Hürlimann et al., 2003; Chen & Lee 2003; Swartz et al., 2003; McArdell et al., 2003; Revellino et al., 2004). The model RAMMS is also based on the Voellmy rheology; the module for debris-flow simulation is available both in a 1D version and in a 2D version (Scheuner et al., 2009; Christen et al., 2012). The latter version of RAMMS was also adapted for the simulation of hillslope debris flows (Christen et al., 2012).

Numerical simulation models applied in case studies to real debris flows include RAMMS (Christen et al., 2010, 2012), DAN or DAN-3D (Ayotte & Hungr 2000; McDougall & Hungr, 2004, 2005), FlatModel (Medina et al., 2008), MassMov2D (Begueria et al., 2009), RASH-3D (Pirulli & Sorbino, 2008), and TRENT-2D (Armanini et al., 2009). Typically, appropriate values for the rheologic or friction parameters were assumed or back-calculated from field observations (Hungr 1995; Rickenmann & Koch 1997; Ayotte & Hungr 2000; Revellino et al., 2004; Naef et al., 2006; Rickenmann et al., 2006; Tecca et al., 2007; Hungr 2008; Hürlimann et al., 2008; Pirulli 2010).

Basically, continuum mechanics simulation models provide the most accurate description of the flow processes, including the deformation of the moving mass along its path as well as detailed spatial and temporal information on the flow parameters. Knowledge of the spatial distribution of the parameters flow velocity and flow depth is important for the production of hazard maps (e.g. BWW/BRP/BUWAL 1997). It must be stressed, however, that generally the rheological model or friction parameters cannot be determined directly (i.e. from samples), but must be assumed on the basis of experience or ideally be "calibrated" from past events in the same region (Rickenmann et al., 2006b; Rickenmann, 2016). Theoretically, based on sediment samples and laboratory tests, the rheological parameters of viscoplastic fluids can be determined for a Bingham or Herschel-Bulkley model; however, this approach typically cannot account for the influence of sediment particles greater than several mm in size.

Various investigations showed that, for the depositional behavior of debris flows on the fan, the topography is a very important and governing factor (Rickenmann et al., 2006b; Scheidl & Rickenmann 2010; Rickenmann & Scheidl 2010). Thus, an appropriate digital terrain model (DTM) must include an accurate representation of the channel and other depressions on the fan For simulations in the context of a hazard assessment, appropriate scenarios for the input conditions have to be defined for example at the fan apex, including assumptions of the total volume of debris flow in an event, the number of surges, and possible depositions on the channel bed due to smaller surges.

4.6 SCENARIOS AND DEPOSITION IN THE AREA OF THE FAN

4.6.1 Uncertainty and scenarios

The term scenario is used here in the context of various assumptions that typically have to be made for the hazard assessment of a future debris-flow event. These assumptions refer, for example, to the number and size of individual surges, the rheologic properties of the mixture (e.g. water content), and possible clogging or log-jam locations that could induce flow overtopping out of the channel. All of these parameters are difficult to predict, yet they can have a big effect on the flow and deposition behavior on the fan by influencing the sequence of flow processes over the duration of a debris-flow event. The establishment of such scenarios is the more challenging the scarcer the information on past events. It is also difficult because the estimation of the frequency of events of a given magnitude is typically imprecise. Thus, one has to work with scenarios that are assigned approximate recurrence periods.

Debris flows generally occur in steep headwater catchments, where often there is a strong interaction between different processes. Hillslope processes can lead to significant sediment delivery to the channels. A larger landslide can rapidly deposit solid material in the channel leading to a (partial) blockage (retention) of the channel discharge with the danger of a collapse of the "dam" thus formed. In steep channels the processes of "flooding", "bedload transport" and "debris flows" often occur in close combination. Here, of interest are not only possible sediment supply sources but, above all, the critical places where an obstruction or complete blockage of the water and sediment discharge may induce an overflowing of the channel banks (e.g. due to a log jam at a bridge cross-section), thereby influencing the potential area affected by debris-flow or bedload deposition.

For the assessment of sediment transfer processes in headwater catchments, relatively large uncertainties still exist with the use of both empirical methods and of numerical simulation models. A proper assessment is also hampered by the difficulty to distinguish between various debris-flow types (granular debris flows, mudflows). Thus, in a technical report on hazard assessment, it is important to state the assumptions made, to point out uncertainties and possibly to perform sensitivity analyses on the chosen input and boundary conditions.

4.6.2 Traces of earlier deposits on the fan

The dominant sediment transfer process (i.e. fluvial bedload transport or debris flow) influences the type of deposit. In the region of debris-flow deposits, all particle sizes are distributed more or less uniformly over the area of deposition, but large boulders at the front of individual debris-flow surges are often transported to the distal end of the deposit. On the other hand, in floods with fluvial bedload transport the coarser components are likely to be deposited on steeper slopes, whereas the finer particles are transported to flatter parts of the fan. In addition, finer particles of debris-flow deposits may be partly re-entrained and re-distributed due to the subsequent runoff that is less sediment-laden.

If there are traces of earlier debris flows on the fan and/or historical documents are available, the assessment of the depositional behavior should be partly based on this information (see also Table 4.2). In comparison with earlier deposits, the size (areas) of the endangered zones may have to be adjusted according to the expected debris load (see also chapter 4.5). Generally steeper fans of irregular fan topography with a rough surface (in the case of a natural fan) point to debris-flow activity. In the evaluation of the traces on the fan (old deposits) primarily the following factors should be taken into account.

4.6.2.1 Old deposits

An accumulation of boulders bigger than about 0.5 m to 1 m diameter points to old debris flow deposits, especially if these deposits lie outside the channel. Levées or debris lobes are (if they are still recognizable) also pointers to areas endangered by debris flow. However, the present fan situation may not necessarily be the same in future. For example, a changed debris-flow topography at the fan apex (after sediment deposition) may induce a new direction of the course of the stream, and, thus, another side of the fan may become endangered. Or a changed future discharge regime with more fluid discharges containing less bedload may lead to a more incised channel at the fan apex, and areas covered earlier by debris may no longer be endangered. Inactive discharge channels and relicts of old stream courses generally point to earlier debris flow activity over the fan, which may have been caused by both debris flows and "normal" flood events with fluvial sediment transport. It should also be noted that due to human use or construction on a fan, old debris traces may no longer be visible.

4.6.2.2 Vegetation cover

If, in the region of the fan, there are areas with tree populations of clearly differing ages, then these may be associated with corresponding debris flow events going back a long time (if a similar influence of snow avalanche events can be excluded).

4.6.2.3 Outcrops in old deposits

If outcrops are present the type of layering may be an indication of earlier debris-flow activity and the individual layer thickness may provide information on the magnitude of the event. If future potential deposition areas are inferred from old debris-flow deposits, possible changes due to channel protection works or new buildings should also be considered. Further, traces of old deposition may have been obliterated by agricultural or other land use on the fan.

4.6.3 Simple assessment of depositional behavior

If there are no traces or indications of earlier events, the potential deposition zones have to be estimated solely using the expected debris load, the topography of the fan, and with assumptions concerning the geometry of the deposition.

An estimate can be made of possible locations of outflow from the channel by comparing the required flow cross-section with the existing channel cross-sections along the fan (see also Fig. 4.13). Hereby, scenarios should also consider, for example, that log jams may form at cross-sections under bridges as a result of driftwood transport (see also chapter 4.6.1). Essentially, the depositional behavior of a debris flow on the fan is controlled by its material properties, in addition to the topography. In the case of a simple estimate of affected areas, first that part of the debris load is determined which will probably be deposited in the channel itself. This part can vary, depending on the location where overflowing on the riverbanks is expected. Even with an adequate channel capacity, filling of the channel is possible; for example, if, in the region of the confluence, there is insufficient room for the deposition of the material or if the receiving stream is not able to transport the material any further. Then, in the second step, the remaining debris load has to be "distributed" over the fan.

If the length of debris-flow deposition on the fan is estimated using empirical approaches, this value should be compared to historical deposits (if any) and if necessary adjusted accordingly. If there is no evidence of old deposits, as a rough approximation it may be assumed that the width of the deposit is about ten times that of the debris-flow width at the fan apex (IKEYA 1981). The mean thickness of the deposit of debris on the fan is often in the range 0.5 m to about 3 m. With these approximate guide values it is possible with some plausibility to distribute the debris load over the fan. Hereby, it is always necessary to keep in mind that the possible overflow locations and the topography as well as structures (buildings, traffic structures) can influence considerably the propagation and depositional behavior.

4.7   FINAL REMARKS

Besides the approaches presented here in chapter 4, other documentations on estimating the various debris-flow parameters can also be found for example in HUNGR et al. (1984), IKEYA (1981, 1987), PWRI (1988), RICKENMANN (1995, 1999), TAKAHASHI (1981), VANDINE (1985, 1996) and HEINIMANN et al. (1998). Regarding the hazard assessment of torrents prone to debris-flow occurrence, mainly several empirical methods have been discussed in detail here. Overall, there is a variety of models to describe the initiation, flow and depositional behavior of debris flows. Each of these models, however, applies to specific material mixtures and boundary conditions. In nature, the material composition and the water content of debris flows can vary greatly. This makes the selection of different types of debris-flow models difficult, because appropriate criteria for that are still largely missing. Therefore empirical approaches continue to have an important part to play.

Regarding the prediction of debris flows, it is possible, in principle, to determine critical values of rainfall for regionally-limited areas, provided that a sufficient number of observations are available concerning past events. However, it is scarcely possible to make a prediction of the exact place (stream channel) of occurrence. Early warning systems are primarily useful to close endangered traffic routes in case of a debris-flow event with the aid of appropriate monitoring devices (CHANG 2003; BADOUX et al., 2009; STÄHLI et al., 2015).

Often the information on earlier debris-flow events is very limited, and estimating the frequency of occurrence of events of a particular magnitude is only possible in approximate terms. Nevertheless, information on earlier debris-flow activity is very important in addition to the assessment of the potential sediment supply during a future event. Such information is not only the basis for the magnitude-frequency relationship, but can also provide important clues for the flow behavior. Therefore it is very important to document past events (e.g. HÜBL et al., 2002) as well as to establish event catalogues of landslides, debris flows and floods.

In engineering practice, the process assessment of debris flows depends mainly on the analogy to earlier events, on an integral assessment of the catchment, on empirical approaches to estimate the most important debris flow parameters, and on simulation models. The most important parameter to estimate the debris-flow hazard is the possible debris load. The use of Geographic Information Systems (GIS) has become more important to estimate the erosion potential and potential debris loads. Simple, semi-empirical models to determine the extent of the debris flow are used for example in combination with GIS, for the preparation of hazard index maps.

Detailed process assessments are necessary for the preparation of hazard maps at a more refined scale. This is often accomplished with numerical models for the simulation of the debris-flow behavior. An uncertainty in this process is the determination or estimation of the model parameters to characterize the material or flow behaviors. In general, these parameters cannot be determined directly using samples of debris material, since the necessary rheological testing equipment for this is not available. Therefore, these parameters have to be estimated by back-calculations of similar events. In assessing the hazards also a possible combination of processes should be taken into account. An example of sediment deposits from a debris flow that were partially eroded again by a subsequent flood on the fan and transported further downstream is illustrated in Fig. 4.21 and Fig. 4.22.

Figure 4.22 Debris-flow deposits of the event of 24 August 1987 on the Minstiger stream fan in Switzerland (photo M. Zimmermann, Thun). The debris flow consisted of a single surge that took place at the beginning of the afternoon, while the fluvial deposits are the result of a flood that occurred in the evening of the same day and also re-entrained finer material from the debris-flow deposits. (Compare also Fig. 4.21).

Chapter 5

Magnitude and frequency of torrent events

Although the relation between the magnitude and the frequency of a debris-flow event is essential for any hazard or risk analysis, it is often difficult to assess. The magnitude of a debris flow event forms an input or basis both for simple empirical relations to estimate important flow parameters (chapter 4.4) and for numerical models simulating flow propagation and deposition (chapter 4.5).

It is very challenging to determine accurately the probability of occurrence of debris flow events of a given magnitude in torrent catchments, because historic data are generally approximate and a detailed assessment of sediment deposits by stratigraphic analysis is typically very expensive (Jakob 2012). This statement is also valid partly also for bedload transporting flood events in torrents. While, in this case, existing rainfall data may facilitate the frequency of rainfall-runoff events of different magnitudes, the sediment supply or sediment availability is much more difficult to assess. If historical data about earlier events are available, they often provide very important information, even if, in general, no statistical evaluation can be made with them in a narrow sense. The traditional concept of extreme value analysis of flood discharges cannot be transferred directly to torrent events; in case of a limited bedload potential, for example, the probability of a future event may largely depend also on the actual stock of movable sediments.

The most important factors in connection with the occurrence of debris flows are the identification of possible triggering zones and sediment supply sources (and, thus, of the event magnitude) as well as the estimation of the frequency of events (Jakob 2005). In torrents prone to debris flow occurrence, the debris load is generally (much) bigger than the sediment load of a flood event with only fluvial bedload transport, and thus is more relevant for the hazard assessment. Here, therefore, the question of appropriate methods to determine the debris load is discussed primarily in detail. The debris loads reported for past events usually include both the volume of solids and pores. In addition, empirical values of debris loads are often based on the solids deposition of a whole event, possibly including multiple debris-flow surges and fluvial bedload transport. In this publication, the event magnitude is assumed to equal the debris load.

To estimate the debris load or the bedload of torrent events the hydrological and geomorphological characteristics of a catchment are important. The sediment supply to the channel network in steep headwater catchments and the total amount of sediment transferred to the fan is controlled both by mass movements on the adjacent hillslopes and erosion and deposition of sediments along the channel (bed and banks) during a flow event (Fig. 5.1).

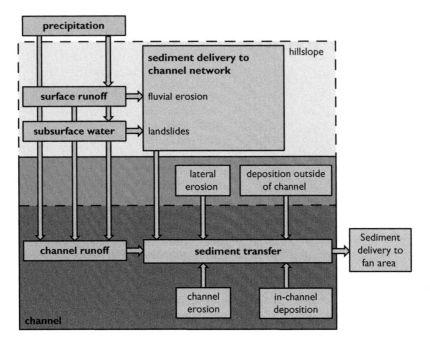

Figure 5.1 Simplified process system for torrents, after LIENER (2000) and GERTSCH (2009). The total amount of sediment transferred to the fan during a torrent event depends on the sediment supply from the hillslopes as well as on the erosion and deposition of sediments along the channel. Modified from LIENER (2000) and GERTSCH (2009).

The fundamental importance of the geologic and geomorphologic aspects of torrent catchments for erosion processes and mass movements is summarized by BUNZA et al. (1976). After STINY (1931), a torrent catchment with relatively young alluvial deposits ("young debris" torrent) is characterized by sediment supply sources consisting mainly of recently weathered rock material. By contrast, according to STINY (1931), in a torrent catchment with abundant residual colluvium ("old debris" torrent) there is generally almost unlimited sediment availability (e.g. due to glacial material consisting of moraine deposits from the various ice ages). The moraine materials, especially those with much fine material, belong to the most hazardous sediment sources of the Alps for torrent floods and debris flows (LUZIAN et al., 2002). The importance of hillslope processes for the sediment supply to channel processes in steep catchments is also discussed for example in BENDA et al. (2005), WICHMANN et al. (2009), GEERTSEMA et al. (2010), and THELER et al. (2010).

In principle, potential future sediment delivery processes may be modeled for entire headwater catchments to estimate possible event magnitudes for example based on rainfall scenarios and slope stability analyses (e.g. MONTGOMERY & DIETRICH, 1994; BAUM et al., 2010; VON RUETTE et al., 2013). However, when applying such models in practical hazard assessment, one of the main challenges is that the characteristics of the soil and subsurface layers are often heterogeneous and unknown. Therefore, the focus here is more on primarily field-based approaches relying on an assessment of hillslope sediment delivery and empirically derived channel erosion rates to be expected during a defined triggering event (in terms of rainfall conditions).

5.1 EMPIRICAL APPROACHES TO ESTIMATE THE MAGNITUDE OF AN EVENT

Some empirical approaches for estimating the debris load or solids load of a torrent event are summarized in Table 5.1. Such approaches usually include simple catchment parameters. They allow an estimate of either an upper limit or of a mean value of the possible debris load or the bedload volume. Only two approaches account for geologic characteristics. In the equation of KRONFELLNER-KRAUS (1984), the coefficient K varies with the geology and the catchment area, with values between about 250 (torrents of the alpine foothills in Austria) and about 1750 (torrents with large sediment sources in residual colluvium [see also chapter 5.5 below]). The value for the geologic-lithologic index I_G after D'AGOSTINO & MARCHI (2001) can take on values in the range 0.5 to 5, depending on the susceptibility to weathering of the surface material.

A comparison between observed bedload as a function of the size of the catchment area exhibits a large scatter of several orders of magnitude (Fig. 5.2); thus, these formulae can only provide very rough estimates. For the development of more reliable approaches, in particular the special geological, geomorphological and hydrological features of a torrent catchment must be taken into account, requiring more detailed investigations.

It may be helpful for the hazard assessment of a particular torrent catchment to compare the estimated event load with the range of values from earlier observations. As an example, for Switzerland there is a compilation of the specific event loads per unit catchment area, grouped according to the predominant geology (i.e. for torrents in the alpine limestone regions, in the crystalline rocks, in the Molasse and Flysch areas) (SPREAFICO et al., 2005; GRASSO et al., 2007) (Fig. 5.3 and Fig. 5.4).

Table 5.1 Simple empirical equations for a rough estimate of the event load of a debris-flow event or a bedload-transporting flood in a torrent; N = number of events as a basis to derive a formula. Definition of the parameters: M = "maximum" event load [m³]; M_a = mean event load [m³]; A_c = catchment area [km²]; S_c = mean channel slope [–]; S_f = mean fan slope [–]; L_c = length of the active channel [m]; K = torrentiality factor; I_G = geologic-lithologic index. (*) This relationship was first derived for event loads in the case of bedload transport, and the coefficients were then adjusted for 15 larger debris flow events in Austria.

Equation	N	Source
$M = K A_c \, 100 \, S_c$	1420	KRONFELLNER-KRAUS (1984); KRONFELLNER-KRAUS (1987)
$M = 27000 \, A_c^{0.78}$	~65	ZELLER (1985); RICKENMANN (1995)
$M_a = 150 \, A_c \, (100 \, S_f - 3)^{2.3}$	15 (*)	HAMPEL (1980)
$M = L_c \, (110 - 250 \, S_f)$	82	RICKENMANN & ZIMMERMANN (1993)
$M_a = 13600 \, A_c^{0.61}$	551	TAKEI (1984)
$M_a = 29100 \, A_c^{0.67}$	64	D'AGOSTINO et al. (1996)
$M_a = 70 \, A_c \, S_c^{1.28} \, I_G$	84	D'AGOSTINO & MARCHI (2001)

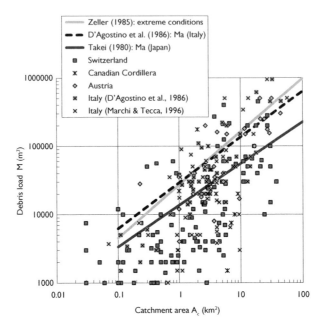

Figure 5.2 Observed event load (magnitude) of debris-flow events, mainly for Switzerland and Northern Italy, as a function of the size of the catchment area. Also shown are some estimating formulae from Table 5.1.

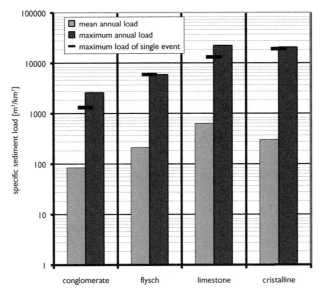

Figure 5.3 Range of observed specific sediment load (normalized by catchment area) subdivided according to geology, based on observed deposition volumes in sediment retention basins in Switzerland. Data from GRASSO et al. (2007) in Hydrological Atlas of Switzerland.

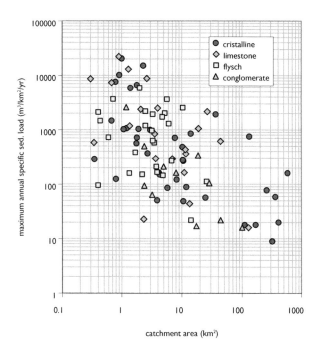

Figure 5.4 Specific annual sediment load (normalized by catchment area) versus catchment area, and subdivided according to geology. There is a tendency for the specific annual event load to decrease with increasing size of the catchment. Data from GRASSO et al. (2007) in Hydrological Atlas of Switzerland.

Further evaluations of these data from sediment retention basins indicate that the shape of the catchment could also have some influence (GRASSO et al., 2010).

5.2 FIELD-BASED ESTIMATE OF THE EVENT MAGNITUDE

For a more exact assessment of the potential event magnitude (debris load or bedload), which could be mobilized during a rainstorm event, a geological-geomorphological assessment of the catchment is performed in many cases, whereby the use of a Geographic Information System (GIS) may be helpful. The main potential triggering areas of debris flows are steep channels or gullies with abundant regolith (colluvium, alluvium) or unstable hillslopes. The latter can also be important for the formation of hillslope debris flows and sediment delivery to the channel network.

A method used frequently in engineering practice to assess the potential event magnitude is to estimate average erosion cross-sections ("channel debris yield rates" according to HUNGR et al., 1984) for more or less homogeneous channel reaches. The resulting erosion volumes are then summed over the whole length of the channel network thought to be affected by sediment entrainment during the event. Typical values of channel debris yield rates as a function of the channel properties and geological-lithological conditions are given in Table 5.2, according to HUNGR et al. (1984). Similar observations on specific channel erosion rates for debris flows and torrent events are reported in SPREAFICO et al. (1996), ZIMMERMANN & LEHMANN (1999), MARCHI & D'AGOSTINO (2004), and HUNGR et al. (2005). According to an investigation of debris-flow events in Switzerland (RICKENMANN & ZIMMERMANN 1993) the mean specific channel erosion rates varied between 40 m³/m and 90 m³/m, and locally values of 500 m² to 650 m² were observed. Such large values were also reported after the outbreak of water from water pockets in glacial areas (HAEBERLI 1983). Outburst flows of water from glacial lakes can lead to very hazardous debris flows, since below the dam breach large discharge peaks can occur and further down-stream there are typically steep channels within morainic material that can often be eroded easily (CLAGUE & EVANS 2000; CHIARLE et al., 2007).

Based on the limited number of observations in Switzerland and Austria an approximate empirical formula was proposed to estimate the "maximum" erosion depth T_e [m] in function of the local channel slope S [m/m] (VAW1992):

$$T_e = 1.5 + 12.5 \, S \tag{5.1}$$

However, as mentioned in chapter 4.3, only very limited field observations are available to document the erosion of debris-flows along the flow path. Therefore, methods for a practical estimation of debris entrainment are largely lacking, despite its importance for the hazard assessment. An example of strong bed and bank erosion along the channel during debris-flow events is shown in Fig. 5.5.

As a method to estimate a potential event magnitude for a torrent catchment based on specific channel erosion rates, in Switzerland the field-based approach of LEHMANN (1993) was further developed (FRICK et al., 2008, 2011; KIENHOLZ et al.,

Table 5.2 Typical values for channel debris yield rates in function of the channel properties and geologic-lithologic conditions from a Canadian investigation of HUNGR et al. (1984). Catchments with areas of 1 to 3 km² were investigated. The stability condition (*) refers to the situation prior to the expected event.

Channel type	Gradient [°]	Bed material	Hillslopes	Stability condition (*)	Channel debris yield rate [m³/m]
A	20–35	bedrock	non-erodible	stable, practically bare of soil cover	0–5
B	10–20	thin debris or loose soil over bedrock	non-e rodible (bedrock)	Stable	5–10
C	10–20	deep talus or moraine	less than 5 m high	Stable	10–15
D	10–20	deep talus or moraine	talus, over 5 m high	hillslopes at repose	15–30
E	10–20	deep talus or moraine	talus, over 20 m high	hillslopes potentially unstable (landslide area)	up to 200 (consider as point source)

Figure 5.5 Strong erosion along the stream channel during the two debris flow events of 18 July and 24 August 1987 in Val Varuna (near Poschiavo, Canton Grisons), Switzerland. (a) Situation before the events (Photo Kraftwerke Brusio AG), (b) Situation after the first event (Photo U. Eggenberger, 29 July 1987), (c) Situation after the second event (Photo G. Paravicini, 29 August 1987). The blue circles mark the positions of an old masonry torrent check dam.

2010). This method called "SEDEX" allows for a more systematic assessment of the sediment contributions of individual channel reaches to estimate a total event volume to be expected at the fan apex. Hereby, several possible event scenarios (e.g. typical rainstorms with a given return period) as well as uncertainties are considered systematically. An important goal of this approach is to ensure the reproducibility and transparency of the assessments.

5.3   COMBINED METHOD FOR ESTIMATING THE EVENT MAGNITUDE

In addition to the method SEDEX, a somewhat more complicated approach for estimating a potential event magnitude for debris flows and fluvial bedload transport in torrents was developed subsequently (GERTSCH 2009; GERTSCH et al., 2010; KIENHOLZ et al., 2010). This method is called "Gertsch" method. In addition to field or map-based estimates of specific channel erosion rates, it includes expert knowledge to modify these first estimated based on a number of catchment characteristics. The method was developed from the analysis of 58 large torrent events in the Swiss Alps (mainly debris flow events) with a recurrence period of at least about 100 years. The method was programmed as a kind of expert system in the form of an Excel file with an assessment of both hillslope and channel processes, and it was validated using 43 test catchments with large torrent events. The method is suitable for catchments smaller than 10 km² in alpine and pre-alpine torrent systems with a mean channel slope well above 10%, and in which active bedload transfer processes can be expected.

The method consists of a system-based approach. It is assumed that the extent of sediment entrainment along a channel reach is given by the characteristics of the given reach, by the conditions and expected processes in the upstream channel reaches, and by threshold processes affecting the entire torrent system as a whole.

The basic catchment parameters can be determined using GIS. Many analysis steps are partly automated and executed in an Excel file. The method does not necessarily require fieldwork, but then the assessment is expected to be less reliable. When using field data additionally, this part of the approach is very similar to the method SEDEX: The results of the method are erosion and deposition loads and thus a sediment budget along the entire flow path. Finally, the expected sediment load at the fan apex can be determined for 100 to 300 year events. A further option is to consider pessimistic scenarios with an assumed return period of more than 300 years.

5.4   FLOOD RUNOFF AND DEBRIS-FLOW OCCURRENCE

For the triggering of debris flows a minimum amount of water is necessary. Debris-flow formation is not only influenced by the surface runoff, which may be largely controlled by rainfall intensity in steep headwater catchments, but also by the degree of soil saturation which is also controlled by rainfall duration. If, through a hillslope instability, a larger sediment volume is moved to a channel, a minimum water input (into the pores of the soil and/or as channel discharge) is required, so that the solids-water mixture is able to reach the fan. Modeling the rainfall-runoff response can be useful in small catchments to estimate a potential water input volume that may limit the maximum emerging debris load that may transform into a debris flow.

Some approaches were proposed to derive a possible debris flow hydrograph based on a pure water hydrograph (GOSTNER et al. 2003). The amount of the entrained solids may then be estimated based on the sediment transport capacity of the water discharge, or more simply by assuming a (constant) bulking factor (GALLINO & PIERSON 1985; PIERSON 1995; BREIEN et al., 2008; GARTNER et al., 2008; SANTI et al., 2008). However, these approaches are subject to large uncertainties. In particular, in this way the maximum discharge of a debris flow can be greatly underestimated, because often the maximum of a debris flow does not correspond just to a simple increase of the peak water discharge by bulking the hydrograph with the additional sediment volume. Estimates show that the maximum discharge of a debris flow may be as much as 10 to 100 greater than the peak discharge of a flood in the same area for the same rainfall conditions (ZIMMERMANN & RICKENMANN 1992; see also Chapter 4 and Table 4.3).

5.5   FREQUENCY OF TORRENT EVENTS

The location and the type of triggering influence primarily the magnitude of the event. The two elements together are important for the flow behavior in the channel. In an area with a limited sediment source potential ("young debris" torrent; see also beginning of chapter 5) the frequency and the magnitude of future debris flows depend on material removal (location, extent) by previous debris flows (ZIMMERMANN et al., 1997a, b).

The most reliable estimate of the possible future frequency of events is based on information of past events. An important source is historical documents (also by spoken communications), while another possibility is in the analysis and interpretation of geomorphological traces. Often the information concerning earlier events is very limited, and the estimate of the frequency of occurrence of events of a certain magnitude is only approximate. Therefore, one has to work with scenarios which are assigned to approximate recurrence periods. A very rough idea, however, can be obtained from the frequency of peak rainfall conditions.

Geomorphological methods to determine debris-flow frequency and, to some extent also, magnitude include dendrochronology (e.g. SCHNEUWLY-BOLLSCHWEILER et al., 2013) and lichenometry (INNES, 2006). Radiocarbon dating may be applied where natural exposures or test trench sediments provide organic materials for dating (CHIVERRELL & JAKOB 2011). Mapping of dated events along with determination of the thickness of the respective deposits can yield magnitude estimations. But even if a comparatively large set of historical data on past debris-flow occurrence is available, the estimation of magnitude-frequency relation with statistical methods may not be straightforward (JAKOB 2012; NOLDE & JOE 2013; RICKENMANN & JAKOB 2015).

5.5.1 Debris-flow events

Characteristic patterns regarding the torrent activity were identified in a study based on 189 historically documented events with mainly debris flows which had occurred in 17 torrent catchments in Switzerland (ZIMMERMANN et al., 1997a, b). These patterns appear to depend mainly on the supply of bedload or the geology and lithology of the catchment. Comparing geomorphological characteristics of the catchments and historical event information resulted in the definition of the following four types of torrent activity (ZIMMERMANN et al., 1997a, b).

1   Torrents with a more or less regular occurrence of debris flows (Fig. 5.6). The time interval of the inactive periods between events amounts to around 15 to 30 years, most bedload is eroded diffusely along the channel, and the sediment load of the whole event is typically smaller than 100,000 m³. The sediment sources frequently consist of relatively young weathered material (mainly "young debris" torrents after STINY 1931), which is eroded along the flow path.

2   Torrents with a rather irregular occurrence of debris flows (Fig. 5.7). After a relatively active period lasting from years to a few decades there may be a longer period of several decades with little to no activity. These torrents run mainly in relatively weak rock of variable strength such as Bündner schist. Debris flow events can transport large amounts of sediments of clearly more than 100,000 m³, which is mainly eroded along the flow path. It may be expected that a major debris flow results in a destabilization of the bed and banks.

3   The occurrence of debris flows is irregular (Fig. 5.8). The catchment is characterized by large abundant debris mainly in moraines and talus slopes ("old debris" torrents after STINY 1931), and the main sediment sources are in the upper part of the catchment. The sediment load is variable and may amount to several 100,000 m³.

4   Torrent catchments for which there are no historical parallels for the occurrence of debris flows (Fig. 5.9). In this category two torrent catchments were identified. An example is the event of 24 August 1987 in the Minstiger stream (Canton

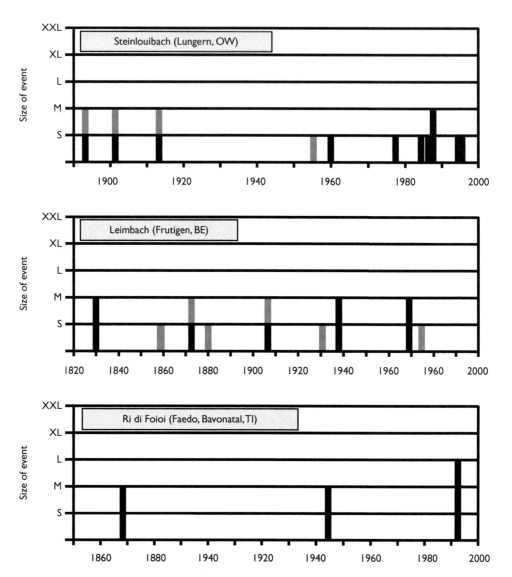

Figure 5.6 Event magnitudes and their frequency in torrent catchments. Here Type I is illustrated for three torrents in Switzerland (ZIMMERMANN et al., 1997a, b). The lighter shaded parts of the bars denote an uncertainty regarding the estimated event magnitude or the event type (possibly flood with bedload transport). Adapted from ZIMMERMANN et al. (1997a, b).

Valais, Switzerland) when a single debris-flow surge reached down to the village of Münster. For a long time, debris-flow activity only occurred in the upper part of the catchment, but for a time period of nearly 300 years before 1987 no similar event has taken place. The event could be related to changes in the subglacial runoff in a warmer climate (ZIMMERMANN et al., 1997a, b).

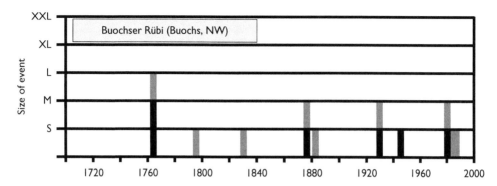

Figure 5.7 Event magnitudes and their frequency in torrent catchments. Here Type 2 is illustrated for one torrent in Switzerland (Zimmermann et al., 1997a, b). The lighter shaded parts of the bars denote an uncertainty regarding the estimated event magnitude or the event type (possibly flood with bedload transport). Adapted from Zimmermann et al. (1997a).

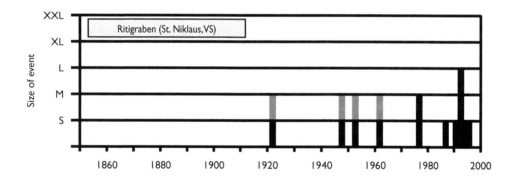

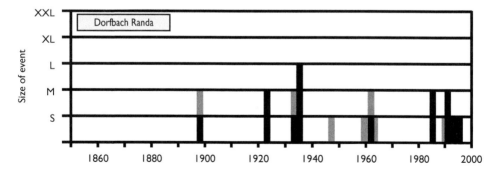

Figure 5.8 Event magnitudes and their frequency in torrent catchments. Here Type 3 is illustrated for two torrents in Switzerland (Zimmermann et al., 1997a, b). The lighter shaded parts of the bars denote an uncertainty regarding the estimated event magnitude or the event type (possibly flood with bedload transport). Adapted from Zimmermann et al. (1997a, b).

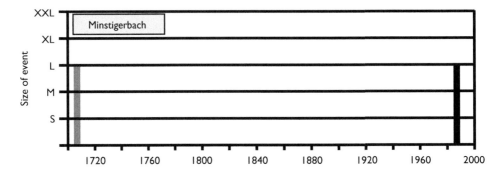

Figure 5.9 Event magnitudes and their frequency in torrent catchments. Here Type 4 is illustrated for one torrent in Switzerland (ZIMMERMANN et al., 1997a, b). The lighter shaded parts of the bars denote an uncertainty regarding the estimated event magnitude or the event type (possibly flood with bedload transport). Adapted from ZIMMERMANN et al. (1997a, b).

5.5.2   Fluvial sediment transport events

For torrent events with fluvial bedload transport (or solids transport), the frequency is determined mostly using the estimated frequency of a flood event or of the corresponding rainfall event. Calculations of fluvial bedload transport are appropriate, for example, in torrent channels with limited bed slopes, where the probability of occurrence of debris flows is small. In this case, estimates of event sediment loads can be made by integrating the bedload transport over the period of a flood hydrograph. With calculations of fluvial bedload transport it has to be taken into account that transport formulae based on total bed shear stress give a maximum solids transport capacity, which is determined from the discharge and the flow hydraulics. However, in steep headwater streams the transported bedload can also be highly dependent on, or limited by, the sediment supply or sediment availability.

5.6   GENERAL REMARKS ON THE ESTIMATE OF SEDIMENT LOADS IN TORRENT CATCHMENTS

The estimation of sediment loads in torrent catchments for events of different return periods is an important task within the framework of a hazard assessment but it is subject to considerable uncertainty. These quantitative approaches are helpful only to a limited extent, and the estimates are based frequently on documented loads from earlier events and are often strongly dependent on expert and field-based evaluations of the conditions in the catchment. Differences in the estimated sediment loads of a factor 2 (between different expert reports) are not uncommon.

A frequent problem is the question to what extent existing or planned protective measures can or should be taken into account in estimating the sediment load and the rainfall-runoff response in the case of torrent catchments. Here too, quantitative statements are often difficult to make. A simpler case is to determine the effect of a sediment retention basin: Basically the expected sediment load downstream can be reduced by the capacity (volume) of the retention basin, if the channel below it is not subject to significant erosion. More difficult to estimate, perhaps, is the influence of a series of torrent check dams on the amount of sediment retained along the channel; this possible sediment retention depends also on the state of the check dams (e.g. type of construction, age, wear and tear or damage) and potential risk of their failure. A detailed discussion of the effectiveness of protection measures in torrent catchments in the context of hazard assessment may be found in PLANAT (2008).

Chapter 6

General remarks on hazard assessment of channel processes in torrents

6.1 REPRODUCIBILITY OF THE PROCESS ANALYSIS AND HAZARD ASSESSMENT

According to KIENHOLZ (1998, 1999, 2002) and PLANAT (2000) the hazard assessment should not only fulfill the requirement of factual correctness, but also guarantee the best possible reproducibility. Securing good reproducibility of this process requires a certain effort regarding the detailed documentation of the assumptions and the methods used in a technical report, but it helps the quality control and simplifies the technical discussion as well as the comparison of different hazard assessments.

The requirement of reproducibility involves ensuring that the selected method of hazard assessment is transparent. In this way, the procedure, the applied approaches and methods, together with the interpretation if the compiled data can be more easily checked. This is also important because an exact evaluation of the process assessment is difficult even after the occurrence of a (larger) event. After KIENHOLZ (1999), therefore, the following basic rules should be adhered to:

- comprehensive documentation (cartographic presentation) of all the relevant perimeters (areas of initiation and of impact of hazard, i.e. the torrent catchment area and the alluvial fan)
- clear choice of methods and their combination and their reporting
- clear-cut decision criteria in the final evaluation
- provision of evidence for identified sub-processes (whose existence can be qualified through traces of earlier events either as "proven", "assumed" or "potential")

6.2 PROCEDURE OF HAZARD ASSESSMENT AND IMPORTANT ASPECTS

In Switzerland, the national platform for natural hazards PLANAT (2000) published recommendations for quality assurance in the assessment of natural hazards. According to this document, the most important elements of the hazard assessment are:

- event documentation/analysis of causes
- map of the phenomena

- hazard map
- risk analysis
- specific hazard assessment for selected critical locations
- protective measures/early warning systems

With regard to and based on the procedure for the hazard assessment, the following sub-steps can be derived. These sub-steps are briefly summarized in connection with the recommendations produced in Switzerland (BWW/BRP/BUWAL 1997; BUWAL/ BWW/BRP 1997):

6.2.1   Basic data

The event documentation and the analysis of causes (especially after large and important events) represent important tasks in the compilation of the basic data. They include investigations of earlier events in the catchment, the dominant processes, the areas affected, the damage observed, and the triggering factors such as the meteorological conditions. A detailed documentation with the analysis of causes also examines how the event evolved and why damage was caused. Potential hazard processes are documented with the map of the phenomena, which can be derived primarily on the basis of characteristics observed in the field and indicators (traces). This map also represents an important part of the fundamentals.

6.2.2   Preparation of hazard map

The main step in the hazard assessment is the preparation of the hazard map. Within a clearly defined perimeter it contains comprehensive information on the hazard potential. The most important elements in the preparation of the hazard map concern, for a torrent catchment, the determination of the dominant hazardous processes (e.g. importance of debris flows, slope instabilities, fluvial bedload transport etc.), the analysis of triggering conditions and the assessment of the probability of occurrence of an event. This is followed by the calculations and modeling of the processes with respect to the spatial spreading out, i.e. especially the process intensities within the potentially endangered areas for a given probability of occurrence. Finally, for each process and the associated event frequency this determines the hazard level (red, blue, yellow, yellow-hatched).

The hazard map serves as the foundation for the risk analysis, the land use planning, the conception and design of protective measures as well as of preventive measures as part of the event management (early warning, emergency planning, etc.). From this perspective it is clear that the creation of a hazard map and the preparation of the corresponding technical report should be as comprehensive as possible and well documented.

The evaluation of the different processes in torrents with a view to the hazard assessment is based mainly on investigations and calculations according to the main steps summarized in this chapter. The most important aspects of this procedure as well as of the sub-processes to be considered are summarized in a compact form in Table 6.1.

Table 6.1 The evaluation of channel processes in torrents and mountain rivers with regard to hazard assessment depends basically on investigations and analyses of the following important aspects or sub-processes.

A. Compilation of basic data
- Aim of the investigation (repair, new construction, planning of hazard zones, etc.)
- Event documentation, chronicle, event register, possibly analysis of triggering processes
- Geomorphologic consideration of the catchment (incl. maps of the processes or maps of the phenomena)
- Geographic parameters of the catchment
- Existing protective measures

B. Magnitude-frequency relationship
- Rainfall (intensity, duration)
- Discharge
- Triggering mechanisms and place of origin in the case of debris flows
- Sediment supply potential/expected sediment load
- Driftwood

To be assessed for different probabilities of occurrence according to BWW/BRP/BUWAL (1997)

Here it should be noted that this step is not independent of steps (C) and (D). In the case of fluvial sediment transport the material entrainment and deposition is typically dependent on the discharge along the channel. With debris flows local initiation zones (e.g. slope instabilities) and sediment inputs in the channel can dominate the total solids transport, but also the material entrainment along the flow path can be very important. Essentially steps (B), (C) and (D) represent an iterative process.

C. Considerations concerning process sequences and scenarios
- Possible Interaction between different processes
- Scenarios in the case of difficult determination of process sequences and of difficult quantification of the corresponding probabilities of occurrence (cf. chapter 4.6)
- Influence of protective measures

D. Analyses of flow, transport and depositional behavior
Important to determine the spatial distribution of the process intensities for given probability of occurrence and event magnitude

D1. Flood formation
- Determination of discharge (peak discharge, hydrograph)
- Hydraulic calculations

D2. Driftwood
- Mobilized load of large wood and dimensions of individual pieces
- Transport capacity
- Possible clogging or log-jam places, identification of endangered areas

D3. Fluvial sediment transport
- Discharge and hydraulics (cf. D1)
- Mobilized volume of solids
- Transport capacity
- Erosion and deposition
- Influence of driftwood (cf. D2)

(Continued)

Table 6.1 (Continued)

D4. Debris flow

- Mobilized volume of solids (slope instabilities and channel erosion)
- Possibly, comparison of expected sediment loads with water runoff volumes
- Solids concentration and sediment properties (flow properties/rheology)
- Debris-flow "hydrograph" at the fan apex
- Flow and depositional behavior
- Influence of driftwood (cf. D2, D3)
- Any process superposition (e.g. fluvial reworking of debris-flow deposits)

E. Uncertainties

- Quantification of uncertainties
- Sensitivity analyses on uncertainties regarding input or model parameters
- Partial consideration of uncertainties by means of scenarios (cf. C)

6.2.2.1　Magnitude-frequency relationship

Determining the probability of occurrence and the corresponding magnitude of an event is often a key step, since essentially in many cases the event magnitude determines the process intensity. Thus it is important to give clear information on how the probability of occurrence of a given process and the corresponding event magnitude (i.e. magnitude-frequency relationship) were assessed.

6.2.2.2　Calculations and process modeling

Another important step is the prediction of the temporal development and spatial displacement of a sub-process ("modeling"), for example, of a debris flow or a flood with bedload transport. Here (numerical) simulation models or, additionally, empirical approaches can be used. To reproduce this sub-process, exact information on the assumptions and basic principles of the calculations are necessary, especially with regard to the selected model parameters. For example, for the modeling of debris flows the selected model parameters should ideally be based on the back-calculation of earlier events. If this is not possible, other assumptions have to be made, e.g. in the use of model parameters as determined for similar areas and material compositions of debris flows based on back-calculations.

6.2.2.3　Formulation of scenarios

The meaning of the term "scenario formulation" as used here (see also chapter 4.6) developed against the background that, above all, the interaction between different processes is often difficult to quantify, particularly in relation to the probability of occurrence of such a combined effect. Examples of scenarios in this sense include the influence of driftwood with regard to a possible (or not occurring) log jam in the case of a bridge cross-section, or the occurrence of a larger landslide supplying sediment to a channel, which depending on the place and type of input can have different effects in the hazard zone. Also generally important for scenario formulation is the investigation of possible critical channel locations (e.g. regarding flow overtopping from channel), since for a given probability of occurrence and event magnitude, depending on the place of channel blockage and outflow, different hazard situations can arise.

6.2.2.4 Uncertainties

The uncertainties should be explicitly mentioned in a technical report and, as far as possible, quantified. This applies especially to the determination of the magnitude-frequency relationship, cf. also (i). As already mentioned, the estimate of the sediment loads in torrents for events of different return periods within the framework of a hazard assessment is associated particularly with considerable uncertainty. Differences in the estimated sediment load by a factor 2 are definitely within the range of uncertainty. Further, the uncertainty should be specified especially in relation to the process modeling and the assumed model parameters (ii), and also in relation to the scenarios (iii).

References

ABRAHAMS, A.D. (2003): Bedload transport equation for sheet flow. Journal of Hydraulic Engineering, 129, 159–163.

ABRAHAMS, A.D., LI, G., KRISHNAN, C., ATKINSON, J.F. (2001): A sediment transport equation for interrill overland flow on rough surfaces. Earth Surface Processes and Landforms, 26, 1443–1459.

ABT, S.R., THORNTON, C.I., SCHOLL, B.A., BENDER, T.R. (2013): Evaluation of Overtopping Riprap Design Relationships. Journal of the American Water Resources Association, 49(4), 923–937.

ANASTASI, G. (1984): Geschiebeanalysen im Felde unter Berücksichtigung von Grobkomponenten. Mitteilungen der Versuchsanstalt für Wasserbau, Hydrologie und Glaziologie, Eidgenössische Technische Hochschule Zürich, Nr. 70, 99p.

ANCEY, C., BAIN, V. (2015): Dynamics of glide avalanches and snow gliding. Reviews of Geophysics, 53, 745–784.

ARMANINI, A., SCOTTON, P. (1993): On the Dynamic Impact of a Debris Flowon Structures. Proceedings of XXV IAHR Congress, Tokyo, Tech. Sess. B III, pp. 203–210.

ARMANINI, A., FRACCAROLLO, L., ROSATTI, G. (2009): Two-dimensional simulation of debris flows in erodible channels. Computers & Geosciences, 35, 993–1006.

ASHMORE, P.E. (1982): Laboratory modelling of gravel braided stream morphology. Earth Surface Processes and Landforms, 7, 201–225.

AULITZKY, H. (1973): Berücksichtigung der Wildbach- und Lawinengefahrengebiete als Grundlage der Raumordnung von Gebirgsländern. In: 100 Jahre Hochschule für Bodenkultur, Band IV, Teil 2, pp. 81–117.

AYOTTE, D., HUNGR, O. (2000): Calibration of a runout prediction model for debris flows and avalanches. In: Wieczorek, G.F. and Naeser, N.D. (eds), Debris-Flow Hazards Mitigation: Mechanics, Prediction, and Assessment, Proceedings 2nd International Conference, Taipei, Taiwan, pp. 505–514. Balkema, Rotterdam.

AZIZ, N.M., SCOTT, D.E. (1989): Experiments on sediment transport in shallow flows in high gradient channels. Hydrological Sciences Journal, 34, 465–478.

BADOUX, A., GRAF, C., RHYNER, J., KUNTNER, R., MCARDELL, B.W. (2009): A debris-flow alarm system for the Alpine Illgraben catchment: design and performance. Natural Hazards, 49(3), 517–539.

BARDOU, E. (2002): Méthodologie de diagnostic et prévision des laves torrentielles sur un bassin versant alpin. Thèse no. 2479, Ecole Polytechnique Fédérale de Lausanne. 187 pp. + annexes.

BARTELT, P., SALM, B., GRUBER, U. (1999): Calculating dense-snow avalanche runout using a Voellmy-fluid model with active/passive longitudinal straining. Journal of Glaciology, 45, 242–254.

BATHURST, J.C. (1985): Flow Resistance Estimation in Mountain Rivers. Journal of Hydraulic Engineering, 111, 625–643.

BATHURST, J.C. (1987): Measuring and modelling bedload transport in channels with coarse bed materials. In K. Richards (ed.), River Channels – Environment and Process, pp. 272–294. Blackwell, Oxford.

BATHURST, J.C. (1993): Flow resistance through the channel network. In K. Beven & M. J. Kirkby (eds), Channel Network Hydrology, pp. 69–98, John Wiley, New York.

BATHURST, J.C. (2007): Effect of coarse surface layer on bed-load transport. Journal of Hydraulic Engineering, 133, 1192–1205.

BATHURST, J.C. (2013): Critical conditions for particle motion in coarse bed materials of nonuniform size distribution. Geomorphology, 197, 170–184.

BATHURST, J.C., GRAF, W.H., CAO, H.H. (1982): Initiation of sediment transport in steep channels with coarse bed material. In: Proc. of Euromech 156, Mechanics of Sediment Transport, pp. 207–213.

BATHURST, J.C., GRAF, W.H., CAO, H.H. (1987): Bed load discharge equations for steep mountain rivers. In Thorne, C.R., Bathurst, J.C., & Hey, R.D. (eds), Sediment transport in gravel bed rivers, pp. 453–477. Wiley, Chichester.

BAUM, R.L., GODT, J.W., SAVAGE, W.Z. (2010): Estimating the timing and location of shallow rainfall-induced landslides using a model for transient, unsaturated infiltration. Journal of Geophysical Research, 115, F03013. doi:10.1029/2009 JF001321.

BEGUERÍA, S., ASCH, T.W.J., MALET, J.-P., GRÖNDAHL, S. (2009): A GIS-based numerical model for simulating the kinematics of mud and debris flows over complex terrain. Natural Hazards and Earth System Sciences, 9, 1897–1909.

BENDA, L., HASSAN, M.A., CHURCH, M., MAY, C.L. (2005): Geomorphology of steepland headwaters: the transition from hillslopes to channels. Journal of the American Water Resources Association, 41, 835–851.

BERGER, C., MCARDELL, B.W., SCHLUNEGGER, F. (2011a): Direct measurement of channel erosion by debris flows, Illgraben, Switzerland. Journal of Geophysical Research – Earth Surface, 116, F01002. doi:10.1029/2010 JF001722.

BERGER, C., MCARDELL, B.W., SCHLUNEGGER, F. (2011b): Sediment transfer patterns at the Illgraben catchment, Switzerland: Implications for the time scales of debris flow activities. Geomorphology, 125, 421–432.

BERTI, M., SIMONI, A. (2007): Prediction of debris flow inundation areas using empirical mobility relationships. Geomorphology, 90, 144–161.

BEZZOLA, G.R. (2002): Fliesswiderstand und Sohlenstabilität natürlicher Gerinne unter besonderer Berücksichtigung des Einflusses der relativen Überdeckung. Mitteilungen der Versuchsanstalt für Wasserbau, Hydrologie und Glaziologie, Eidgenössische Technische Hochschule Zürich, Nr. 173, 258p.

BEZZOLA, G.R. (2005): Vorlesungsmanuskript Flussbau, Fassung Wintersemester 2005/2006. Professur für Wasserbau an der Versuchsanstalt für Wasserbau, Hydrologie und Glaziologie, Eidgenössische Technische Hochschule Zürich.

BEZZOLA, G.R., GANTENBEIN, S., HOLLENSTEIN, R., MINOR, H.-E. (2002): Verklausung von Brückenquerschnitten. Int. Symposium Moderne Methoden und Konzepte im Wasserbau. In: Mitteilungen der Versuchsanstalt für Wasserbau, Hydrologie und Glaziologie, Eidgenössische Technische Hochschule Zürich, Nr. 175, pp. 87–97.

BEZZOLA, G.R., SCHILLING, M., OPLATKA, M. (1996): Reduzierte Hochwassersicherheit durch Geschiebe. Schweizer Ingenieur und Architekt, 41, 886–892.

BRAUDRICK, C., GRANT, G.E. (2000): When do logs move in rivers. Water Resources Research, 36, 571–584.

BRAUDRICK, C., GRANT, G.E. (2001): Transport and deposition of large woody debris in streams: A flume experiment. Geomorphology, 41, 263–284.

BRAUNER, M. (1999): Modelling the sediment budget of an alpine catchment within a GIS environment. In: Proceedings of the 28th IAHR Congress (1999), Graz, CD-ROM, 6p.

BRAUNER, M. (2001): Aufbau eines Expertensystems zur Erstellung einer ereignisbezogenen Feststoffbilanz in einem Wildbacheinzugsgebiet. Dissertation am Institut für Alpine Naturgefahren und Forstliches Ingenieurwesen, Universität für Bodenkultur, Wien.

BRAY, D.I. (1979): Estimating average velocity in gravel-bed rivers. Journal of the Hydraulic Division, 105(HY9), 1103–1122.

BREIEN, H., DE BLASIO, F.V., ELVERHØI, A., HØEG, K. (2008): Erosion and morphology of a debris flow caused by a glacial lake outburst flood, western Norway. Landslides, 5, 271–280.

BUFFINGTON, J.M., MONTGOMERY, D.R. (1997): A systematic analysis of eight decades of incipient motion studies, with special reference to gravel-bed rivers. Water Resources Research, 33, 1993–2029.

BUNTE, K., ABT, S.R. (2001): Sampling surface and subsurface particle size distributions in wadable gravel- and cobble-bed streams for analysis in sediment transport, hydraulics, and streambed monitoring. U.S. Dep. of Agric., For. Serv., Rocky Mt. Res. Stn., Fort Collins, Colorado, USA.Gen. Tech. Rep. RMRS-GTR-74, 428p.

BUNTE, K., ABT, S.R., SWINGLE, K.W., CENDERELLI, D.A., SCHNEIDER, J.M. (2013): Critical Shields values in coarse-bedded steep streams. Water Resources Research, 49, 1–21. doi:10.1002/2012 WR012672.

BUNZA, G., KARL, J., MANGELSDORF, J. (1982): Geologisch-morphologische Grundlagen der Wildbachkunde. Schriftenreihe des Bayerischen Landesamtes für Wasserwirtschaft, Heft 17, München, 128p.

BUSCOMBE, D., RUBIN, D.M., WARRICK, J.A. (2010): A universal approximation of grain size from images of noncohesive sediment. Journal of Geophysical Research – Earth Surface, 115, F02015. doi:10.1029/2009JF001477.

BUWAL/BWW/BRP (1997): Empfehlungen: Berücksichtigung der Massenbewegungsgefahren bei raumwirksamen Tätigkeiten. Herausgeber: Bundesamt für Umwelt, Wald und Landschaft (BUWAL), Bundesamt für Wasserwirtschaft (BWW), Bundesamt für Raumplanung (BRP), Bern, 42p.

BWW/BRP/BUWAL (1997): Empfehlungen: Berücksichtigung der Hochwassergefahren bei raumwirksamen Tätigkeiten. Herausgeber: Bundesamt für Wasserwirtschaft (BWW), Bundesamt für Raumplanung (BRP), Bundesamt für Umwelt, Wald und Landschaft (BUWAL), Biel, 32p.

CAINE, N. (1980): The rainfall intensity-duration control of shallow landslides and debris flows. Geografiska Annaler, 62 A, 23–27.

CANNON, S.H. (1993): An empirical model for the volume- change behavior of debris flows. In: H.W. Shen, S.T. Su & F. Wen (eds), Hydraulic Engineering 93,Vol. 2, American Society of Civil Engineers, New York, pp. 1768–1773.

CARSON, M.A. (1987): Measures of flow intensity as predictors of bed load. Journal of Hydraulic Engineering, 113, 1402–1421.

CAO, Z., PENDER, G., WALLIS, S., CARLING, P. (2004): Computational dam-break hydraulics over erodible sediment bed. Journal of Hydraulic Engineering, 130, 689–703.

CHANG, S.Y. (2003): Evaluation of a system for detecting debris flows and warning road traffic at bridges susceptible to debris flow hazard. In: D. Rickenmann & C.L. Chen (eds), Debris-Flow Hazards Mitigation: Mechanics, Prediction, and Assessment, pp. 731–742. Millpress, Rotterdam.

CHANSON, H. (2004): Hydraulics of Open Channel Flow. 2nd Edition, Elsevier, 585p.

CHEN, H., LEE, C.F. (2003): A dynamic model for rainfall-induced landslides on natural slopes. Geomorphology, 51, 269–288.

CHIARI, M., RICKENMANN, D. (2009): Modellierung des Geschiebetransportes mit dem Modell SETRAC für das Hochwasser im August (2005) in Schweizer Gebirgsflüssen. Wasser, Energie, Luft, 101, 319–327.

CHIARI, M., RICKENMANN, D. (2011): Back-calculation of bedload transport in steep channels with a numerical model. Earth Surface Processes and Landforms, 36, 805–815.

CHIARI, M., SCHEIDL, C. (2015): Application of a new cellular model for bedload transporting extreme events at steep slopes. Geomorphology, 246, 413–419.

CHIARI, M., FRIEDL, K., RICKENMANN, D. (2010): A one dimensional bedload transport model for steep slopes. Journal of Hydraulic Research, 48, 152–160.

CHIARLE, M., IANNOTTI, S., MORTARA, G., DELINE, P. (2007): Recent debris flow occurrences associated with glaciers in the Alps. Global and Planetary Change, 56, 123–36.

CHIEW, Y.M., PARKER, G. (1994): Incipient sediment motion on non-horizontal slopes. Journal of Hydraulic Research, 32(5), 649–660.

CHIN, A. (1989): Step–pools in stream channels. Progress in Physical Geography, 13, 391–408.

CHIVERRELL, R., JAKOB, M. (2011): Radiocarbon dating: alluvial fan/debris cone evolution and hazards. In: M. Schneuwly-Bollschweiler, M. Stoffel, F. Rudolf-Miklau (eds), Dating torrential processes on fans and cones - methods and their application for hazard and risk assessment; Advances in Global Change Research, 47, 265–282.

CHOI, S.U., GARCIA, M.H. (1993): Kinematic wave approximation for debris flow routing. Proc. XXV IAHR Congress, Technical Session B, Vol. III, pp. 94–101.

CHRISTEN, M., BARTELT, P., KOWALSKI, J. (2010): Back calculation of the In den Arelen avalanche with RAMMS: interpretation of model results. Annals of Glaciology, 51, 161–168.

CHRISTEN, M., BÜHLER, Y., BARTELT, P., LEINE, R., GLOVER, J., SCHWEIZER, A., GRAF, C., MCARDELL, B.W., GERBER, W., DEUBELBEISS, Y., FEISTL, T., VOLKWEIN, A. (2012): Integral hazard management using a unified software environment: numerical simulation tool "RAMMS" for gravitational natural hazards. In: Proc. International Symposium Interpraevent, Grenoble, France, Vol. 1, pp. 77–86.

CHURCH, M. (2013): Steep Headwater Channels. In: Shroder, J., Wohl, E. (eds), Treatise on Geomorphology. Academic Press, San Diego, CA, vol. 9, Fluvial Geomorphology, pp. 528–549.

CHURCH, M., MCLEAN, D., WOLCOTT, J.F. (1987): River bed gravels: Sampling and analysis. In Thorne, C.R., Bathurst, J.C., Hey, R.W. (eds), Sediment transport in gravel bed rivers, Wiley, Chichester, pp. 43–79.

CLAGUE, J.J., EVANS, S.G. (2000): A review of catastrophic drainage of moraine-dammed lakes in British Columbia. Quaternary Science Reviews, 19, 1763–1783.

COMITI, F., MAO, L. (2012): Recent advances on the dynamics of steep channels. In: Church, M., Biron, P., Roy, A.G. (eds), Gravel Bed Rivers 7: Processes, Tools, Environments. Wiley-Blackwell, Chichester, pp. 353–377.

COMITI, F., D'AGOSTINO, V.D., MOSER, M., LENZI, M.A., BETTELLA, F., AGNESE, A.D., RIGON, E., GIUS, S., MAZZORANA, B. (2012): Preventing wood-related hazards in mountain basins: from wood load estimation to designing retention structures. Proceedings, 12th Congress INTERPRAEVENT 2012, Grenoble, France; 651–662.

COROMINAS, J. (1996): The angle of reach as a mobility index for small and large landslides. Canadian Geotechnical Journal, 33, 260–271.

COSTA, J.E. (1984): Physical geomorphology of debris flows. In: J.E. Costa & P.J. Fleischer (eds), Developments and applications of geomorphology, Springer, Berlin, 268–317.

COSTA, J.E. (1988): Rheologic, geomorphic, and sedimentologic differentiation of water floods, hyperconcentrated flows, and debris flows. In: V.R. Baker, R.C. Kochel & P.C. Patton (eds), Flood Geomorphology, John Wiley & Sons, New York, pp. 113–122.

COUSSOT P. (1997): Mudflow rehology and dynamics. IAHR Mongraph Series. Rotterdam: Balkema. 255p.

COUSSOT, P., LAIGLE, D., ARATTANO, M., DEGANUTTI, A., MARCHI, L. (1998): Direct determination of rheological characteristics of debris flow. J Hydraul Eng, 124, 865–868.

Cui, P., Zeng, C., Lei, Y. (2015): Experimental analysis on the impact force of viscous debris flow. Earth Surface Processes and Landforms, 40, 1644–1655.

David, G.C.L., Wohl, E.E., Yochum, S.E., Bledsoe, B.P. (2010): Controls on spatial variations in flow resistance along steep mountain streams. Water Resources Research, 46, W03513. doi:10.1029/2009WR008134.

D'Agostino, V., Cerato M., Coali, R. (1996): Il trasporto solido di eventi estremi nei torrenti del Trentino Orientale. [Sediment transport of extreme events in torrents of eastern Trentino], Proc. International Symposium Interpraevent, Garmisch-Partenkirchen, Germany, Bd. 1, pp. 377–386 [in Italian].

D'Agostino, V., Marchi, L. (2001): Debris flow magnitude in the Eastern Italian Alps: data collection and analysis. Physics and Chemistry of the Earth (C), 26, 657–663.

De Jong, C., Ergenzinger, P. (1995): The interrelations between mountain valley form and river-bed arrangement. In: E.J. Hickin (ed.), River Geomorphology, pp. 55–91. John Wiley & Sons Ltd.

Detert M., Weitbrecht V. (2012a): Automatic object detection to analyze the geometry of gravel grains – a free stand-alone tool. In: Proc. of the 6th International Conference on Fluvial Hydraulics, River Flow 2012, San José (Costa Rica), pp. 595–600. Boca Raton, London.

Detert M., Weitbrecht V. (2012b): BASEGRAIN 1.0. Wasser, Energie, Luft, 104, 334.

Dittrich, A. (1998): Wechselwirkung Morphologie/Strömung naturnaher Fliessgewässer. Mitteilungen des Institutes für Wasserwirtschaft und Kulturtechnik, Universität Karlsruhe, Heft 198, 208p.

Eaton, B.C. (2013): Hydraulic geometry: empirical investigations and theoretical approaches. In: Shroder, J., Wohl, E. (eds), Treatise on Geomorphology. Academic Press, San Diego, CA, vol. 9, Fluvial Geomorphology, pp. 313–329.

Egashira, S., Ashida, K. (1991): Flow resistance and sediment transportation in streams with step-pool bed morphology. In: Fluvial Hydraulics of Mountain Regions, pp. 45–58. Springer: Heidelberg.

Egashira, S., Honda, N., Itoh, T. (2001): Experimental study on the entrainment of bed material into debris flow. Physics and Chemistry of the Earth, Part C, 26(9), 645–650.

Egli, T. (2005): Wegleitung Objektschutz gegen gravitative Naturgefahren. Vereinigung Kantonaler Feuerversicherungen VKF, Bern, Kapitel 5 Murgänge, pp. 77–87.

Einstein, A. (1942): Formulas for the transportation of bed load. Transactions of the American Society of Civil Engineers, 107, 561–597.

Fannin, R.J., Wise, M.P. (2001): An empirical-statistical model for debris flow travel distance. Canadian Geotechnical Journal, 38, 982–994.

Fehr, R. (1986): A method for sampling very coarse sediments in order to reduce scale effects in movable bed models. In: Proc. Symposium on Scale effects in modelling sediment transport phenomena, Toronto, IAHR, Delft, pp. 383–397.

Fehr, R. (1987a): Geschiebeanalysen in Gebirgsflüssen – Umrechnung und Vergleich von verschiedenen Analyseverfahren. Mitteilungen der Versuchsanstalt für Wasserbau, Hydrologie und Glaziologie, Eidgenössische Technische Hochschule Zürich, Nr. 92, 139p.

Fehr, R. (1987b): Einfache Bestimmung der Korngrössenverteilung von Geschiebematerial mit Hilfe der Linienzahlanalyse. Schweizer Ingenieur und Architekt, 38, 1104–1109.

Ferguson, R. (2007): Flow resistance equations for gravel- and boulder-bed streams. Water Resources Research, 43, W05427. doi:10.1029/(2006)WR005422.

Fraccarollo, L., Papa, M. (2000): Numerical Simulation of Real Debris-Flow Events. Physics and Chemistry of the Earth, Part B: Hydrology, Oceans and Atmosphere, 25(9), 757–763.

French, R.H., Miller, J.J., Curtis, S. (2001): Estimating the depth of deposition (erosion) at slope transitions on alluvial fans. Journal of Hydraulic Engineering, 127, 780–782.

FRICK, E., KIENHOLZ, H., ROTH, H. (2008): SEDEX – eine Methodik zur gut dokumentierten Abschätzung der Feststofflieferung in Wildbächen. Wasser, Energie, Luft, 100, 131–136.

FRICK, E., KIENHOLZ, H., ROMANG, H. (2011): SEDEX (SEDiments and EXperts), Anwender-Handbuch. Geographica Bernensia P42, Geographisches Institut der Universität Bern, 128p., ISBN 978-3-905835-27-4.

FUCHS, S., KAITNA, R., SCHEIDL, S., HÜBL, J. (2008): The Application of the Risk Concept to Debris Flow Hazards. Geomechanik und Tunnelbau, 1(2), 120–129.

FULLER, W., THOMPSON, S.E. (1907): The laws of proportioning concrete. Transactions of the American Society of Civil Engineers, 59, 67–143.

GALLINO, G.L., PIERSON, T.C. (1985): Polallie Creek debris flow and subsequent dam-break flood of 1980. East Fork Hood River Basin. Oregon. U.S. Geological Survey, Water-Supply Paper, 2273, 22pp.

GAMMA, P. (2000): dfwalk – Ein Murgangsimulationprogramm zur Gefahrenzonierung. Geographica Bernensia, G66, Geographisches Institut der Universität Bern, 144p.

GARBRECHT, G. (1961): Abflussberechnung für Flüsse & Kanäle. Die Wasserwirtschaft, 51, (2): 40–45, (4): 72–77.

GARTNER, J.E., CANNON, S.H., SANTI, P.M., deWOLFE, V.G. (2008): Models to predict debris flow volumes generated by recently burned basins. Geomorphology, 96, 339–354.

GAUCKLER, P.G. (1867): Etudes Théoriques et Pratiques sur l'Ecoulement et le Mouvement des Eaux. Comptes Rendues de l'Academie des Sciences, Paris, France, Tome 64, pp. 818–822.

GEERTSEMA, M., SCHWAB, J.W., JORDAN, P., MILLARD, T.H., ROLLERSON, T.P. (2010): Hillslope Processes. In R.G. Pike et al. (editors), Compendium of forest hydrology and geomorphology in British Columbia. B.C. Min. For. and Range, For. Sci. Prog., Victoria, B.C. and FORREX Forum for Research and Extension in Natural Resources, Kamloops, B.C. Land Management Handbook 66, pp. 213–273.

GENEVOIS, R., TECCA, P.R., BERTI, M., SIMONI, A. (2000): Debris-flow in the Dolomites: Experimental data from a monitoring system. In: G.F. Wieczorek & N.D. Naeser (eds), Debris-Flow Hazards Mitigation: Mechanics, Prediction, and Assessment; Proceedings 2nd International DFHM Conference, Taipei, Taiwan, August 16–18, 2000, pp. 283–291. Rotterdam: Balkema.

GENOLET, F. (2002): Modélisation de laves torrentielles – Contribution à la paramétrisation du modèle Voellmy-Perla. Postgraduate thesis, Ecole Polytechnique Fédérale de Lausanne, Suisse, 70p.

GEO (2000): Review of natural terrain landslide debris-resisting barrier design. Geotechnical Engineering Office (GEO), Civil Engineering Department, The Government of the Hong Kong Special Administrative Region, Special Project Report SPR 1/(2000).

GEORGE, D.L., IVERSON, R.M. (2014): A depth-averaged debris-flow model that includes the effects of evolving dilatancy: II. Numerical predictions and experimental tests. Proc. R. Soc. Lond. Ser. A470, 20130820. http://dx.doi.org/10.1098/rspa.2013.0820.

GERTSCH, E. (2009): Geschiebelieferung alpiner Wildbachsysteme bei Grossereignissen – Ereignisanalysen und Entwicklung eines Abschätzverfahrens. Dissertation am Geographischen Institut der Universität Bern, 203p. [http://www.zb.unibe.ch/download/eldiss/09 gertsch_e.pdf]

GERTSCH, E., KIENHOLZ, H., SPREAFICO, M. (2010): Projektbericht. Geschiebelieferung alpiner Wildbachsysteme bei Grossereignissen – Ereignisanalysen und Entwicklung eines Abschätzverfahrens. Hydrologie und Wasserbewirtschaftung, 54, 310–315.

GOMEZ, B., CHURCH, M. (1989): An assessment of bedload sediment transport formulae for gravel bed rivers. Water Resources Research, 25, 1161–1186.

GOSTNER, W., BEZZOLA, G.R., SCHATZMANN, M., MINOR, H.E. (2003): Integral analysis of debris flow in Alpine torrent – the case study of Tschengls. In: D. Rickenmann & C.L. Chen (eds), Debris-Flow Hazards Mitigation: Mechanics, Prediction, and Assessment, Proceedings of 3rd International DFHM Conference, Davos, Switzerland, September 10–12, 2003, pp. 1129–1140. Rotterdam: Millpress.

GRAF, W.H., SUSZKA, L. (1987): Sediment transport in steep channels. Journal of Hydroscience and Hydraulic Engineering, 5, 11–26.

GRAHAM, D.J., ROLLET, A.J., PIEGAY, H., RICE, S.P. (2010): Maximizing the accuracy of image-based surface sediment sampling techniques. Water Resources Research, 46, W02508. doi:10.1029/2008WR006940.

GRASSO, A., DOBMANN, J., JAKOB, A. (2007): Hydrological Atlas of Switzerland, Plate 7.8: Bed-Material Loads in Selected Catchments.

GRASSO, A., JAKOB, A., SPREAFICO, M., BÉROD, D. (2010): Monitoring von Feststofffrachten in schweizerischen Wildbächen. Wasser, Energie, Luft, 102, 41–45.

GRANT, G.E., SWANSON, F.J., WOLMAN, M.G. (1990): Pattern and origin of stepped-bed morphology in high gradient streams, Western Cascades, Oregon. Geological Society of America Bulletin, 102, 340–352.

GREMINGER, P. (2003): Managing the risks of natural hazards. In D. Rickenmann & C.L. Chen (eds), 3rd Int. Conf. on Debris-Flow Hazards Mitigation. Millpress, Rotterdam, The Netherlands, pp. 39–56.

GRIFFITHS, G.A. (2003): Downstream hydraulic geometry and hydraulic similitude. Water Resources Research, 39(4), 1094. doi: 10.1029/2002 WR001485.

GÜNTER, A. (1971): Die kritische mittlere Sohlenschubspannung bei Geschiebemischungen unter Berücksichtigung der Deckschichtbildung und der turbulenzbedingten Sohlenschubspannungsschwankungen. Mitteilungen der Versuchsanstalt für Wasserbau, Hydrologie und Glaziologie, Eidgenössische Technische Hochschule Zürich, Nr. 3, 69p.

GURNELL, A.M. (2013): Wood in fluvial systems. In: Shroder, J., Wohl, E. (eds), Treatise on Geomorphology. Academic Press, San Diego, CA, vol. 9, Fluvial Geomorphology, pp. 163–188.

GUZZETTI, F., PERUCCACCI, S., ROSSI, M., STARK, C.P. (2007): Rainfall thresholds for the initiation of landslides. Meteorology and Atmospheric Physics, 98, 239–267.

GUZZETTI, F., PERUCCACCI, S., ROSSI, M., STARK, C.P. (2008): The rainfall intensity–duration control of shallow landslides and debris flows: an update. Landslides, 5, 3–17.

HAEBERLI, W. (1983): Frequency and characteristics of glacier floods in the Swiss Alps. Annals of Glaciology, 4, 85–90.

HAEBERLI, W., RICKENMANN, D., ZIMMERMANN, M. (1991): Murgänge. Ursachenanalyse der Unwetterereignisse (1987): Mitteilungen des Bundesamtes für Wasserwirtschaft, Bern, Schweiz, Nr. 4, pp. 77–87.

HAGER, W.H. (2001): Gauckler and the GMS formula. Journal of Hydraulic Engineering, 127(8), 635–638.

HAMPEL, R. (1980): Grundlagen für Gefahrenzonen in Wildbächen. Proc. International Symposium Interpraevent, Bad Ischl, Austria, Bd. 3, pp. 83–91.

HAN, G., WANG, D. (1996): Numerical modeling of Anhui debris flow. Journal of Hydraulic Engineering, 122(5), 262–265.

HARTLIEB, A., BEZZOLA, G.R. (2000): Ein Überblick zur Schwemmholzproblematik. Wasser, Energie, Luft, 92, 1–5.

HASSAN, M.A., CHURCH, M., LISLE, T.E. (2005a): Sediment transport and channel morphology of small, forested streams. Journal of the American Water Resources Association, 41, 853–876.

HASSAN, M.A., HOGAN, D.L, BIRD, S.A, MAY, C.L., GOMI T., CAMPBEL, D. (2005b): Spatial and temporal dynamics of wood in headwater streams of the pacific northwest. Journal of the American Water Resources Association, 41(4), 899–919.

HAYWARD, J.A. (1980): Hydrology and stream sediments in a mountain catchment, in Tussock Grasslands and Mountain Lands Institute Special Publ. no. 17 (Ph. D. dissertation, Lincoln College, Canterbury), New Zealand, 235p.

HEGG, C., RICKENMANN, D. (1999): Comparison of bedload transport in a steep mountain torrent with a bedload transport formula. In: Hydraulic Engineering for Sustainable Water Resources Management at the Turn of the Millenium. Proceedings 28th IAHR Congress, 22–27 August (1999) in Graz, Austria. [CD-ROM] Graz, Technical University, 7p.

HEIMANN, F.U.M., RICKENMANN, D., TUROWSKI, J.M., KIRCHNER, J.W. (2015a): sedFlow – a tool for simulating fractional bedload transport and longitudinal profile evolution in mountain streams. Earth Surface Dynamics, 3, 15–34.

HEIMANN, F.U.M., RICKENMANN, D., BÖCKLI, M., BADOUX, A., TUROWSKI, J.M., KIRCHNER, J.W. (2015b): Calculation of bedload transport in Swiss mountain rivers using the model sedFlow: proof of concept. Earth Surface Dynamics, 3, 35–54.

HEINIMANN, H.R., HOLLENSTEIN, K., KIENHOLZ, H., KRUMMENACHER, B., MANI, P. (1998): Methoden zur Analyse und Bewertung von Naturgefahren. Umwelt-Materialien Nr. 85. Bundesamt für Umwelt, Wald und Landschaft, Bern, 248p.

HEY, R.D. (1979): Flow Resistance in Gravel Bed Rivers. Journal of the Hydraulics Division, 105 (HY4), 365–379.

HODEL, H. (1993): Untersuchung zur Geomorphologie, der Rauheit des Strömungswiderstandes und des Fliessvorganges in Bergbächen. Dissertation Nr. 9830, Eidgenössische Technische Hochschule Zürich, 289p.

HOFMEISTER, R.J., MILLER, D.J. (2003): GIS-based modeling of debris-flow initiation, transport and deposition zones for regional hazard assessments in western Oregon, USA. In: D. Rickenmann & C.L. Chen (eds), Debris-Flow Hazards Mitigation: Mechanics, Prediction, and Assessment, Proceedings 3rd International DFHM Conference, Davos, Switzerland, pp. 1141–1149. Rotterdam: Millpress.

HÜBL, J. (2006): Vorläufige Erkenntnisse aus 1:1 Murenversuchen: Prozessverständnis und Belastungsannahmen. In: FFIG, G. Reiser (Hrsg.), Geotechnik und Naturgefahren: Balanceakt zwischen Kostendruck und Notwendigkeit. Institut für Geotechnik, BOKU Universität Wien, Geotechnik und Naturgefahren, 19.10.2006, Wien.

HÜBL, J., KIENHOLZ, H., LOIPERSBERGER, A. (eds) (2002): DOMODIS-Documentation of Mountain Disasters, State of Discussion in the European Mountain Areas. International Research Society Interpraevent, Klagenfurt, Austria. http://wasser.ktn.gv.at/interpraevent.

HUNGR, O. (1995): A model for the runout analysis of rapid flow slides, debris flows, and avalanches. Canadian Geotechnical Journal, 32, 610–623.

HUNGR, O. (2008): Numerical Modelling of the Dynamics of Debris Flows and Rock Avalanches. Geomechanik und Tunnelbau, 1(2), 112–119.

HUNGR, O., MCDOUGALL, S. (2009): Two numerical models for landslide dynamic analysis. Computers & Geosciences, 35, 978–992.

HUNGR, O., MORGAN, G.C., KELLERHALS, R. (1984): Quantitative analysis of debris torrent hazards for design of remedial measures. Canadian Geotechechnical Journal, 21, 663–677.

HUNGR, O., MCDOUGALL, S., BOVIS, M. (2005): Entrainment of material by debris flows. In: M. Jakob & O. Hungr (eds), Debris-Flow Hazards and Related Phenomena, pp. 135–158. Heidelberg: Praxis-Springer.

HUNGR O., Leroueil, S., Picarelli, L. (2014): The Varnes classification of landslide types, an update. Landslides 11, 167–194.

HUNGR, O., EVANS, S.G., BOVIS, M.J., HUTCHINSON, J.N. (2001): A review of the classification of landslides of the flow type. Environmental & Engineering Geoscience, 7, 221–238.

HUNZIKER, R.P. (1995): Fraktionsweiser Geschiebetransport. Mitteilungen der Versuchsanstalt für Wasserbau, Hydrologie und Glaziologie, Eidgenössische Technische Hochschule Zürich, 138, 191p.

HUNZIKER, R.P., JAEGGI, M.N.R. (2002): Grain sorting processes. Journal of Hydraulic Engineering, 128, 1060–1068.

HUNZINGER, L., ZARN, B. (1996): Geschiebetransport und Ablagerungsprozesse in Wildbachschalen. Proc. International Symposium Interpraevent, Garmisch-Partenkirchen, Germany, Bd. 4, pp. 221–230.

HÜRLIMANN, M., RICKENMANN, D., GRAF, C. (2003): Field and monitoring data of debris-flow events in the Swiss Alps. Canadian Geotechechnical Journal, 40, 161–175.

HÜRLIMANN, M., McArdell, B.W., Rickli, C. (2015): Field and laboratory analysis of the runout characteristics of hillslope debris flows in Switzerland. Geomorphology, 232, 20–32.

HÜRLIMANN, M., RICKENMANN, D., MEDINA, V., BATEMAN, A. (2008): Evaluation of approaches to calculate debris-flow parameters for hazard assessment. Engineering Geology, 102, 152–163.

IKEYA, H. (1979): Introduction to SABO works. The Japan Sabo Association, Tokyo, First English Edition, 1979, 168pp.

IKEYA, H. (1981): A method of designation for area in danger of debris flow. In: Erosion and Sediment Transport in Pacific Rim Steeplands, International Association of Hydrological Sciences, Publ. no. 132, pp. 576–588.

IKEYA, H. (1987): Debris flow and its countermeasures in Japan. Bulletin International Association of Engineering Geologists, 40, 15–33.

IMRAN, J., PARKER, G., LOCAT, J., LEE, H. (2001): 1D Numerical model of muddy subaqueous and subaerial debris flows. Journal of Hydraulic Engineering, 127(11), 959–967.

INNES, J.L. (2006): Lichenometric dating of debris-flow deposits in the Scottish Highlands. Earth Surface Processes and Landforms, 8, 579–588.

IVERSON, R.M. (2012): Elementary theory of bed-sediment entrainment by debris flows and avalanches. Journal of Geophysical Research, 117, F03006. doi:10.1029/2011JF002189.

IVERSON, R.M., DENLINGER, R.P. (2001): Flow of variably fluidized granular masses across three-dimensional terrain, 1. Coulomb mixture theory. Journal of Geophysical Research, 106(B1), 537–552.

IVERSON, R.M., GEORGE, D.L. (2014): A depth-averaged debris-flow model that includes the effects of evolving dilatancy: I. Physical basis. Proc. R. Soc. Lond. Ser. A470, 20130819. http://dx.doi.org/10.1098/rspa.2013.0819.

IVERSON, R.M., SCHILLING, S.P., VALLANCE, J.W. (1998): Objective delineation of lahar-inundation zones. Geological Society of America Bulletin, 110, 972–984.

IVERSON, R.M., GEORGE, D.L., ALLSTADT, K., REID, M.E., COLLINS, B.D., VALLANCE, J.W., SCHILLING, S.P., GODT, J.W., CANNON, C.M., MAGIRL, C.S., BAUM, R.L., COE, J.A., SCHULZ, W.H., BOWER, J.B. (2015): Landslide mobility and hazards: implications of the 2014 Oso disaster. Earth and Planetary Science Letters, 420, 197–208. doi: 10.1016/j.epsl.2014.12.020.

JACKSON, W.L., BESCHTA, R.L. (1982): A model of two-phase bedload transport in an Oregon Coast Range stream. Earth Surface Processes and Landforms, 7, 517–527.

JACKSON, K.J., WOHL, E. (2015): Instream wood loads in montane forest streams of the Colorado Front Range, USA. Geomorphology, 234, 161–170.

JÄGGI, M.N.R. (1992): Sedimenthaushalt und Stabilität von Flussbauten. Mitteilungen der Versuchsanstalt für Wasserbau, Hydrologie und Glaziologie, Eidgenössische Technische Hochschule Zürich, 119, 100p.

JAKOB, M. (2005): Debris-flow hazard analysis. In Jakob, M., Hungr, O. (eds), Debris-flow hazards and related phenomena, Praxis and Springer, Heidelberg, pp. 411–443.

JAKOB, M. (2012): The fallacy of frequency – Statistical techniques for debris-flow frequency-magnitude analyses. In: E. Eberhardt, C.A. Froese, K. Turner & S. Leroueil (eds), Landslides and Engineered Slopes: Protecting society through improved understanding, pp. 741–750. CRC Press/Balkema.

JARRETT, R.D. (1984): Hydraulics of high-gradient streams. Journal of Hydraulic Engineering, 110, 1519–1539.

JIN, M., FREAD, D.L. (1999): 1D modeling of mud/debris unsteady flows. Journal of Hydraulic Engineering, 125(8), 827–834.

JULIEN, P.Y., O'BRIEN, J.S. (1997): On the importance of mud and debris flow rheology in structural design. In C.L. Chen (ed.), Debris-Flow Hazards Mitigation: Mechanics, prediction, and Assessment, pp. 350–359. New York: ASCE.

KAITNA, R. (2006): Debris flow experiments in a rotating drum. Dissertation, Institut für Alpine Naturgefahren, Universität für Bodenkultur Wien, 170p.

Kaitna, R., Chiari, M., Kerschbaumer, M., Kapeller, H., Zlatic-Jugovic, J., Hengl, M., Huebl, J. (2011): Physical and numerical modelling of a bedload deposition area for an Alpine torrent, Natural Hazards and Earth System Sciences, 11, 1589–1597.

Kasprak, A., Magilligan, F.J., Nislow, K.H., Snyder, N.P. (2012): A LIDAR derived evaluation of watershed-scale large woody debris sources and recruitment mechanisms: coastal Maine, USA. River Research and Applications, 28: 1462–1476.

Katul, G., Wiberg, P., Albertson, J., Hornberger, G. (2002): A mixing layer theory for flow resistance in shallow streams. Water Resources Research, 38, 1250. doi:10.1029/(2001)WR000817.

Kellerhals, R., Bray, D.I. (1971): Sampling Procedures for Coarse Fluvial Sediments. Proceedings American Society of Civil Engineers, Journal of the Hydraulics Division, 97(HY8), 1165–1979.

Keulegan, G.H. (1938): Laws of turbulent flow in open channels. Journal of Research of the National Bureau of Standards, Vol. 21, Research Paper 1151, 707–741.

Kienholz, H. (1999): Anmerkungen zur Beurteilung von Naturgefahren in den Alpen. In: Relief, Boden, Paläoklima, Vol. 14, pp. 165–184. Berlin und Stuttgart: Gebr. Borntraeger.

Kienholz, H., Herzog, B., Bischoff, A., Willi, H.P. (2002): Fragen der Qualitätssicherung bei der Gefahrenbeurteilung. Bündnerwald, 55, 57–67.

Kienholz, H., Frick, E., Gertsch, E. (2010): Assessment tools for mountain torrents: SEDEX© and bed load assessment matrix. Proc. International Interpraevent Symposium, Taipei, Taiwan, pp. 245–256.

Kronfellner-Krauss, G. (1984): Extreme Feststofffrachten und Grabenbildungen von Wildbächen. Proc. International Symposium Interpraevent, Villach, Austria, Bd. 2, 109–118.

Kronfellner-Krauss, G. (1987): Zur Anwendung der Schätzformel für extreme Wildbach-Feststofffrachten im Süden und Osten Oesterreichs. Wildbach- und Lawinenverbau, 51, 187–200.

Laigle, D., Coussot, P. (1997): Numerical modelling of debris flows. Journal of Hydraulic Engineering 123, 617–623.

Lamb, M.P., Dietrich, W.E., Venditti, J.G. (2008): Is the critical Shields stress for incipient sediment motion dependent on channel-bed slope? Journal of Geophysical Research – Earth Surface, 113, F0(2008). doi:10.1029/(2007)JF000831.

Lange, D., Bezzola G.R. (2006): Schwemmholz, Probleme und Lösungsansätze. Mitteilungen der Versuchsanstalt für Wasserbau, Hydrologie und Glaziologie, Eidgenössische Technische Hochschule Zürich, Nr. 188, 125p.

Lassettre, N.S., Kondolf, G.M. (2012): Large woody debris in urban stream channels: redefining the problem. River Research and Application, 28, 1477–1487.

Legros, F. (2002): The mobility of long-runout landslides. Engineering Geology, 63, 301–331.

Lehmann, C. (1993): Zur Abschätzung der Feststofffracht in Wildbächen – Grundlagen und Anleitung. Geographica Bernensia G42, Bern.

Lenzi, M., Mao, L., Comiti, F. (2004): Magnitude-frequency analysis of bed load data in an Alpine boulder bed stream. Water Resources Research, 40, W07201. doi:10.1029/2003WR002961.

Liener, S. (2000): Zur Feststofflieferung in Wildbächen. Geographica Bernensia G64, Bern.

Lucía, A., Comiti, F., Borga, M., Cavalli, M., Marchi, L. (2015): Dynamics of large wood during a flash flood in two mountain catchments. Natural Hazards and Earth System Sciences, 15, 1741–1755.

Luzian, R., Kohl, B., Bichler, I., Kohl, J., Bauer, W. (2002): Wildbäche und Muren – Eine Wildbachkunde mit einer Übersicht von Schutzmassnahmen der Ära Aulitzky. Forstliche Bundesversuchsanstalt, Wien, ISBN 3-901347-34-8, 163p.

Mächler, D. (2009): GIS-Modellierung von potentiellen Schwemmholzeinträgen durch Rutschungen. Semesterarbeit, Umweltingenieurwesen, Zürcher Hochschule für angewandte Wissenschaften (ZHAW), Wädenswil, 22 p.

MALET, J.P., MAQUAIRE, O., LOCAT, J., REMAÎTRE, A. (2004): Assessing debris flow hazards associated with slow moving landslides: methodology and numerical analyses. Landslides, 1, 83–90.

MANNING, R. (1891): On the flow of water in open channels and pipes. Transactions of the Institution of Civil Engineers of Ireland, 20, 161–207.

MARCHI, L., TECCA, P.R. (1996): Magnitudo delle colate detritiche nelle Alpi Orientali Italiane. Geoingegneria Ambientale e Mineraria, 33(2/3), 79–86 (in Italian).

MARCHI, L., BROCHOT, S. (2000): Les cônes de déjection torrentielles dans les Alpes françaises; morphométrie et processus de transport solide torrentiel. Revue de géographie alpine, 88, 23–38.

MARCHI, L., ARATTANO, M. & DEGANUTTI, A.M. (2002): Ten years of debris flows monitoring in the Moscardo Torrent (Italian Alps): Geomorphology, 46(1/2), 1–17.

MARCHI, L., D'AGOSTINO, V. (2004): Estimation of debris-flow magnitude in the Eastern Italian Alps. Earth Surf. Process. Landforms, 29(2), 207–220.

MATHYS, N., BROCHOT, S., MEUNIER, M., RICHARD, D. (2003): Erosion quantification in the small marly experimental catchments of Draix (Alpes de Haute Provence, France): Calibration of the ETC rainfall-runoff-erosion model. Catena, 50, 527–548.

MAZZORANA, B., ZISCHG, A., LARGIADER, A., HÜBL, J. (2009): Hazard index maps for woody material recruitment and transport in alpine catchments. Natural Hazards and Earth System Science, 9, 197–209.

MAZZORANA, B., COMITI, F., VOLCAN, C., SCHERER, C. (2011): Determining flood hazard patterns through a combined stochastic–deterministic approach. Natural Hazards, 59(1), 301–316.

McARDELL, B.W., ZANUTTIGH, B., LAMBERTI, A., RICKENMANN, D. (2003): Systematic comparison of debris flow laws at the Illgraben torrent, Switzerland. In D. Rickenmann & C.L. Chen (eds), Debris-Flow Hazards Mitigation: Mechanics, Prediction, and Assessment; Proceedings 3rd International DFHM Conference, Davos, Switzerland, September 10–12, 2003, pp. 647–657. Rotterdam: Millpress.

McARDELL, B.W., BARTELT, P., KOWALSKI, J. (2007): Field observations of basal forces and fluid pore pressure in a debris flow. Geophysical Research Letters, 34, L07406. doi:10.1029/(2006)GL029183.

McCOY, S.W., KEAN, J.W., COE, J.A., TUCKER, G.E., STALEY, D.M., WASKLEWICZ, T.A. (2012): Sediment entrainment by debris flows: In situ measurements from the headwaters of a steep catchment. Journal of Geophysical Research, 117, F03016. doi:10.1029/2011JF002278.

McDOUGALL, S., HUNGR, O. (2004): A model for the analysis of rapid landslide motion across three-dimensional terrain. Canadian Geotechnical Journal, 41, 1084–1097.

McDOUGALL, S., HUNGR, O. (2005): Dynamic modelling of entrainment in rapid landslides. Canadian Geotechnical Journal, 42, 1437–1448.

MEDINA, V., HÜRLIMANN, M., BATEMAN, A. (2008): Application of FLAT Model, a 2D finite volume code, to debris flows in the northeastern part of the Iberian Peninsula. Landslides, 5, 127–142.

MEYER-PETER, E., MÜLLER, R. (1948): Formulas for bedload transport. In: Proceedings 2nd meeting Int. Assoc. Hydraulic Structures Res., Stockholm, Sweden, Appendix 2, pp. 39–64.

MEYER-PETER, E., MÜLLER, R. (1949): Eine Formel zur Berechnung des Geschiebetriebs. Schweizerische Bauzeitung, 67, 29–32.

MIZUYAMA, T. (1981): An intermediate phenomenon between debris flow and bed load transport. In: Erosion and Sediment Transport in Pacific Rim Steeplands, International Association of Hydrological Sciences, Publ. no. 132, pp. 212–224.

MIZUYAMA, T., KOBASHI, S., OU, G. (1992): Prediction of debris flow peak discharge. Proc. International Symposium Interpraevent, Bern, Switzerland, Bd. 4, pp. 99–108.

MONTGOMERY, D.R., DIETRICH, W.E. (1994): A physically based model for the topographic control on shallow landsliding. Water Resources Research, 30(4), 1153–1171.

MONTGOMERY, D.R., BUFFINGTON, J.M. (1997): Channel reach morphology in mountain drainage basins. Geological Society of America Bulletin, 109, 596–611.

MORVAN, H., KNIGHT, D., WRIGHT, N., TANG, X., CROSSLEY, A. (2008): The concept of roughness in fluvial hydraulics and its formulation in 1D, 2D and 3D numerical simulation models, Journal of Hydraulic Research, 46, 191–208.

NÄF, D., RICKENMANN, D., RUTSCHMANN, P., MCARDELL, B.W. (2006): Comparison of flow resistance relations for debris flows using a one-dimensional finite element simulation model. Natural Hazards and Earth System Sciences, 6, 155–165.

NAKAGAWA, H., TAKAHASHI, T., SATOFUKA, Y. (2000): A debris-flow disaster on the fan of the Harihara River, Japan. In G.F. Wieczorek & N.D. Naeser (eds), Debris-Flow Hazards Mitigation: Mechanics, Prediction, and Assessment; Proceedings 2nd International DFHM Conference, Taipei, Taiwan, August 16–18, 2000, pp. 193–201. Rotterdam: Balkema.

NAKAMURA, J. (1980): Investigation manual on prediction of occurrence of dosekiryu [debris flows], delineation of dangerous zone affected by dosekiryu and arrangement of warning and evacuation system in mountain torrents in Japan. Proc. International Symposium Interpraevent, Bad Ischl, Austria, Bd. 3, pp. 41–81.

NAMEGHI, A.E., HASSANLI, A.M., SOUFI, M. (2008): A study of the influential factors affecting the slopes of deposited sediments behind the porous check dams and model development for prediction. DESERT, 12, 113–119. (http://jdesert.ut.ac.ir)

NITSCHE, M., TUROWSKI, J.M., BADOUX, A., PAULI, M., SCHNEIDER, J., RICKENMANN, D., KOHOUTEK, T.K. (2010): Measuring streambed morphology using range imaging. In: A. Dittrich, K. Koll, J. Aberle & P. Geisenhainer (eds), River Flow 2010, Bundesanstalt für Wasserbau, pp. 1715–1722. ISBN 978-3-939230-00-7.

NITSCHE, M., RICKENMANN, D., TUROWSKI, J.M., BADOUX, A., KIRCHNER, J.W. (2011): Evaluation of bedload transport predictions using flow resistance equations to account for macro-roughness in steep mountain streams. Water Resources Research, 47, W08513. doi:10.1029/2011WR010645.

NITSCHE, M., RICKENMANN, D., KIRCHNER, J.W., TUROWSKI, J.M., BADOUX, A. (2012a): Macro-roughness and variations in reach-averaged flow resistance in steep mountain streams, Water Resources Research, 48, W12518. doi: 10.1029/2012WR012091.

NITSCHE, M., RICKENMANN, D., TUROWSKI, J.M., BADOUX, A., KIRCHNER, J.W. (2012b): Verbesserung von Geschiebevorhersagen in Wildbächen und Gebirgsflüssen durch Berücksichtigung von Makrorauigkeit. Wasser, Energie, Luft, 104, 129–139.

NNADI, F.N., WILSON, K.C. (1992): Motion of contact-load particles at high shear stress. Journal of Hydraulic Engineering, 118, 1670–1684.

NOLDE, N., JOE, H. (2013): A Bayesian extreme value analysis of debris flows. Water Resources Research, 49, 7009–7022.

O'BRIEN, J.S., JULIEN, P.Y., FULLERTON, W.T. (1993): Two-dimensional water flood and mudflow simulation. Journal of Hydraulic Engineering, 119, 244–261.

PAGLIARA, S., CHIAVACCINI, P. (2006): Flow Resistance of Rock Chutes with Protruding Boulders. Journal of Hydraulic Engineering, 132, 545–552.

PALT, S.M. (2001): Sedimenttransporte im Himalaya-Karakorum und ihre Bedeutung für Wasserkraftanlagen. Mitteilungen des Institutes für Wasserwirtschaft und Kulturtechnik der Universität Karlsruhe, Heft 209, 257p.

PALT, S.M., DITTRICH, A. (2002): Stabilität von Gebirgsflüssen und rauen Rampen. Österreichische Wasser- und Abfallwirtschaft, 54, 75–86.

PARKER, G., WILCOCK, P.R., PAOLA, C., DIETRICH, W.E., PITLICK, J. (2007): Physical basis for quasi-universal relations describing bankfull hydraulic geometry of single-thread gravel bed rivers. Journal of Geophysical Research, 112, F04005. doi:10.1029/2006JF000549.

PARKER, G. (2008): Transport of gravel and sediment mixtures. In: M.H. Garcia (ed.), Sedimentation engineering: Theories, measurements, modeling, and practice, pp. 165–252. Reston, USA: ASCE.

PETRASCHECK, A., KIENHOLZ, H. (2003): Hazard assessment and mapping of mountain risks in Switzerland. In D. Rickenmann & C.L. Chen (eds), 3rd Int. Conf. on Debris-Flow Hazards Mitigation. Millpress, Rotterdam, The Netherlands, pp. 25–38.

PHILLIPS, C.J., DAVIES, T.R.H. (1991): Determining rheological parameters of debris flow material. Geomorphology, 4, 101–110.

PIERSON, T.C. (1986): Flow behavior of channelized debris flows, Mount St. Helens, Washington. In: A.D. Abrahams (ed.), Hillslope Processes, pp. 269–296. Boston, USA: Allen and Unwin.

PIERSON, T.C. (1995): Flow characteristics of large eruption-triggered debris flows at snow-clad volcanoes: constraints for debris-flow models. Journal of Volcanology and Geothermal Research, 66, 283–294.

PIERSON, T. (2005): Distinguishing between debris flows and floods from field evidence in small watersheds. USGS Fact Sheet 2004–3142, January 2005.

PIERSON, T.C., COSTA, J.E. (1987): A rheologic classification of subarerial sediment-water flows. Geological Society of America, Reviews in Engineering Geology, Vol. VII, 1–12.

PIRULLI, M. (2010): On the use of the calibration-based approach for debris-flow forward-analyses. Natural Hazards and Earth System Sciences, 10, 1009–1019.

PIRULLI, M., SORBINO, G. (2008): Assessing potential debris flow runout: a comparison of two simulation models. Natural Hazards and Earth System Sciences, 8, 961–971.

PITON, G., RECKING, A. (2015a): Design of sediment traps with open check dams. I: Hydraulic and deposition processes. Journal of Hydraulic Engineering. doi:10.1061/(ASCE)HY.1943–7900.0001048, 04015045.

PITON, G., RECKING, A. (2015b): Design of Sediment Traps with Open Check Dams. II: Woody Debris. Journal of Hydraulic Engineering. doi: 10.1061/(ASCE)HY.1943–7900.0001049.

PLANAT (2000): Empfehlungen zur Qualitätssicherung bei der Beurteilung von Naturgefahren. Nationale Plattform Naturgefahren (PLANAT), Bern, 20p.

PLANAT (2008): Wirkung von Schutzmassnahmen. Nationale Plattform Naturgefahren (PLANAT), Bern, Strategie Naturgefahren Schweiz, Umsetzung des Aktionsplans PLANAT 2005–2008, Projekt A3, Schlussbericht 2. Phase, Testversion, Dezember 2008. 289p.

PORTO, P., GESSLER, J. (1999): Ultimate bed slope in Calabrian streams upstream of check dams: field study. Journal of Hydraulic Engineering, 125, 1231–1242.

PROCHASKA, A.B., SANTI, P.M., HIGGINS, J., CANNON, S.H. (2008): Debris-flow runout predictions based on the average channel slope (ACS). Engineering Geology, 98, 29–40.

PWRI (1988): Technical standard for measures against debris flow (Draft). Technical Memorandum of Public Works Research Institute (PWRI), No. 2632, Ministry of Construction, Japan, 48p.

RAETZO, H., RICKLI, C. (2007): Rutschungen. In: G.R. Bezzola & C. Hegg (eds), Ereignisanalyse Hochwasser (2005), Teil 1 – Prozesse, Schäden und erste Einordnung. Bundesamt für Umwelt BafU, Bern und Eidg. Forschungsanstalt für Wald, Schnee und Landschaft WSL, Birmensdorf, pp. 195–209.

RECKING, A. (2009): Theoretical development on the effects of changing flow hydraulics on incipient bed load motion. Water Resources Research, 45, W04401. doi:10.1029/(2008)WR006826.

RECKING, A. (2013): An analysis of nonlinearity effects on bed load transport prediction. Journal of Geophysical Research – Earth Surface, 118, 1264–1281. doi:10.1002/jgrf.20090.

REID, L.M., DUNNE, T. (1996): Rapid evaluation of sediment budgets. Geo-Ecology Texts, Catena Verlag, Reiskirchen, Germany, 164p.

REVELLINO, P., HUNGR, O., GUADAGNO, F.M., EVANS, S.G. (2004): Velocity and runout simulation of destructive debris flows and debris avalanches in pyroclastic deposits, Campania region, Italy. Environmental Geology, 45, 295–311.

RICKENMANN, D. (1990): Bedload transport capacity of slurry flows at steep slopes. Mitteilungen der Versuchsanstalt für Wasserbau, Hydrologie und Glaziologie, Eidgenössische Technische Hochschule Zürich, 103, 249p.

RICKENMANN, D. (1991): Hyperconcentrated flow and sediment transport at steep slopes. Journal of Hydraulic Engineering, 117(11), 1419–1439.

RICKENMANN, D. (1994): An alternative equation for the mean velocity in gravel-bed rivers and mountain torrents. In G.V. Cotroneo & R.R. Rumer (eds), Proceedings ASCE 1994 National Conference on Hydraulic Engineering, Buffalo N.Y., USA, Vol. 1, pp. 672–676.

RICKENMANN, D. (1995): Beurteilung von Murgängen. Schweizer Ingenieur und Architekt, 48, 1104–1108.

RICKENMANN, D. (1996): Fliessgeschwindigkeit in Wildbächen und Gebirgsflüssen. Wasser, Energie, Luft, 88, 298–304.

RICKENMANN, D. (1996): Murgänge: Prozess, Modellierung und Gefahrenbeurteilung. In B. Oddsson (ed.), Instabile Hänge und andere risikorelevante natürliche Prozesse, Nachdiplomkurs in angewandten Erdwissenschaften, pp. 397–407. Basel: Birkhäuser.

RICKENMANN, D. (1997a): Sediment transport in Swiss torrents. Earth Surface Processes and Landforms, 22, 937–951.

RICKENMANN, D. (1997b): Schwemmholz und Hochwasser. Wasser, Energie, Luft, 89, 115–119.

RICKENMANN, D. (1999): Empirical relationships for debris flows. Natural Hazards, 19, 47–77.

RICKENMANN, D. (2001a): Comparison of bed load transport in torrents and gravel bed streams. Water Resources Research, 37, 3295–3305.

RICKENMANN, D. (2001b): Murgänge in den Alpen und Methoden zur Gefahrenbeurteilung. In: Mitteilungen des Lehrstuhls und Instituts für Wasserbau und Wasserwirtschaft, Rheinisch-Westfälische Technische Hochschule Aachen, Nr. 124, pp. 51–77.

RICKENMANN, D. (2005a): Geschiebetransport bei steilen Gefällen. In: Mitteilungen der Versuchsanstalt für Wasserbau, Hydrologie und Glaziologie, Eidgenössische Technische Hochschule Zürich, Nr. 190, pp. 107–119.

RICKENMANN, D. (2005b): Runout prediction methods. In: M. Jakob & O. Hungr (eds), Debris-Flow Hazards and Related Phenomena, pp. 263–282. Heidelberg: Praxis-Springer.

RICKENMANN, D. (2008): Lastfälle aus Murgangprozessen – Bemessungsgrundlagen. Herbstkurs der Fachleute für Naturgefahren (FAN), Bellinzona, Schweiz, 17.9.2008.

RICKENMANN, D. (2012): Alluvial steep channels: flow resistance, bedload transport and transition to debris flows. In: M. Church, P. Biron & A. Roy (eds), Gravel Bed Rivers: Processes, Tools, Environment, pp. 386–397. Chichester, England: John Wiley & Sons.

RICKENMANN, D. (2016): Debris-flow hazard assessment and methods applied in engineering practice. International Journal of Erosion Control Engineering (Japan), in press (http://www.jsece.or.jp/jece/index.html).

RICKENMANN, D., ZIMMERMANN, M. (1993): The 1987 debris flows in Switzerland: documentation and analysis. Geomorphology, 8, 175–189.

RICKENMANN, D., KOCH, T. (1997): Comparison of debris flow modelling approaches. In C.L. Chen (ed.), Debris-Flow Hazards Mitigation: Mechanics, Prediction, and Assessment, Proceedings 1st International DFHM Conference, San Francisco, CA, USA, August 7–9, 1997, pp. 576–585. New York: ASCE.

RICKENMANN, D., WEBER, D. (2000): Flow resistance of natural and experimental debris flows in torrent channels. In G.F. Wieczorek & N.D. Naeser (eds), Debris-Flow Hazards Mitigation: Mechanics, Prediction, and Assessment, Proceedings 2nd International DFHM Conference, Taipei, Taiwan, August 16–18, 2000, pp. 245–254. Rotterdam: Balkema.

RICKENMANN, D., MCARDELL, B.W. (2007): Continuous measurement of sediment transport in the Erlenbach stream using piezoelectric bedload impact sensors. Earth Surface Processes and Landforms, 32, 1362–1378.

RICKENMANN, D., SCHEIDL, C. (2010): Modelle zur Abschätzung des Ablagerungsverhaltens von Murgängen. Wasser, Energie, Luft, 102, 17–26.

RICKENMANN, D., KOSCHNI, A. (2010): Sediment loads due to fluvial transport and debris flows during the 2005 flood events in Switzerland. Hydrological Processes, 24, 993–1007.

RICKENMANN, D., RECKING, A. (2011): Evaluation of flow resistance equations using a large field data base. Water Resources Research, 47, W07538. doi:10.1029/2010WR009793.

RICKENMANN, D., JAKOB, M. (2015): Erosion and sediment flux in mountain watersheds. In: The High-Mountain Cryosphere, C. Huggel, M. Carey, J.J. Clague & A. Kääb (eds), Cambridge University Press, pp. 166–183.

RICKENMANN, D., HÜRLIMANN, M., GRAF, C., NÄF, D., WEBER, D. (2001): Murgang-Beobachtungsstationen in der Schweiz. Wasser, Energie, Luft, 93, 1–8.

RICKENMANN, D., WEBER, D., STEPANOV, B. (2003): Erosion by debris flows in field and laboratory experiments. In: D. Rickenmann & C.L. Chen (eds), Debris-Flow Hazards Mitigation: Mechanics, Prediction, and Assessment, Proceedings of 3rd International DFHM Conference, Davos, Switzerland, September 10–12, 2003, pp. 883–894. Rotterdam: Millpress.

RICKENMANN, CHIARI, M., FRIEDL, K. (2006a): SETRAC – A sediment routing model for steep torrent channels. In R. Ferreira, E. Alves, J. Leal & A. Cardoso (eds), River Flow 2006, pp. 843–852. London: Taylor & Francis.

RICKENMANN, D., LAIGLE, D., MCARDELL, B.W., HÜBL, J. (2006b): Comparison of 2D debris-flow simulation models with field events. Computational Geosciences, 10, 241–264.

RICKENMANN, D., HEIMANN, F.U.M., BÖCKLI, M. TUROWSKI, J.M., BIELER, C., BADOUX, A. (2014): Geschiebetransport-Simulationen mit sedFlow in zwei Gebirgsflüssen der Schweiz. Wasser Energie Luft (eingereicht).

RICKENMANN, D., BÖCKLI, M., HEIMANN, F.U.M., BADOUX, A., TUROWSKI, J.M. (2015): Das Modell sedFlow und Erfahrungen aus Simulationen des Geschiebetransportes in fünf Gebirgsflüssen der Schweiz. Synthesebericht. WSL Berichte, Heft 24, 68p. (www.wsl.ch/publikationen/pdf/14594.pdf).

RICKLI, C., BUCHER, H.U. (2006): Einfluss ufernaher Bestockungen auf das Schwemmholzvorkommen in Wildbächen. Projektbericht zuhanden des Bundesamtes für Umwelt BAFU. Eidgenössische Forschungsanstalt WSL, Birmensdorf, Schweiz, 94p.

RICKLI, C., RAETZO, H., MCARDELL, B., PRESLER, J. (2008): Hanginstabilitäten. In: G.R. Bezzola & C. Hegg (eds), Ereignisanalyse Hochwasser 2005: Teil 2 – Analyse von Prozessen, Massnahmen und Gefahrengrundlagen. Bundesamt für Umwelt BAFU, Bern, Eidgenössische Forschungsanstalt WSL, Birmensdorf, pp. 97–116.

RIMBÖCK, A. (2003): Schwemmolzrückhalt in Wildbächen. Grundlagen zu Planung und Berechnung von Seilnetzsperren. Berichte des Lehrstuhls und der Versuchsanstalt für Wasserbau und Wasserwirtschaft, Nr. 94, Technische Universität München, 163p.

RINDERER, M., JENEWEIN, S., SENFTER, S., RICKENMANN, D., SCHÖBERL, F., STÖTTER, J., HEGG, C. (2009): Runoff and bedload transport modelling for flood hazard assessment in small alpine catchments – the model PROMAB-GIS. In E. Veulliet, J. Stötter & H. Weck-Hannemann (eds), Sustainability in Natural Hazard Management, pp. 69–101. Berlin: Springer-Verlag.

ROMANG, H. (2004): Wirksamkeit und Kosten von Wildbach-Schutzmassnahmen. Geographica Bernensia, G 73, Universität Bern, ISBN 3-906151-76-X, 211pp.

ROSPORT, M. (1998): Fliesswiderstand und Sohlstabilität steiler Fliessgewässer unter Berücksichtigung gebirgsbachtypischer Sohlstrukturen. Mitteilungen des Institutes für Wasserwirtschaft und Kulturtechnik der Universität Karlsruhe, Heft 196, 144p.

RÖSSERT, R. (1978): Hydraulik im Wasserbau. 4. Auflage, Oldenbourg, München.

RUDOLF-MIKLAU, F. (2001): Untersuchungen an kohäsionslosen Sedimenten in kalkalpinen Wildbächen der Steiermark (Österreich). Dissertation, Universität für Bodenkultur Wien.

RUDOLF-MIKLAU F., HÜBL J., SCHATTAUER G., RAUCH H. P., KOGELNIG A., HABERSACK H., SCHULEV-STEINDL E. (2011): Handbuch Wildholz – Praxisleitfaden. Internationale Forschungsgesellschaft Interpraevent, Klagenfurt, 32p.

RUF, G. (1990): Fliessgeschwindigkeiten in der Ruetz/Stubaital/Tirol. Wildbach- und Lawinenverbau, 54(115), 219–227.

RUIZ-VILLANUEVA, V., BLADÉ, E., DÍEZ-HERRERO, A., BODOQUE, J.M., SÁNCHEZ-JUNI, M. (2014): Two dimensional modelling of large wood transport during flash floods. Earth surface processes and landforms, 39, 438–449.

RYAN, S.E., PORTH, L.S., TROENDLE, C.A. (2005): Coarse sediment transport in mountain streams in Colorado and Wyoming, USA. Earth Surface Processes and Landforms, 30, 269–288.

SANTI, P.M., deWOLFE, V.A., HIGGINS, J.D., CANNON, S.H., GARTNER, J.E., SANTI, P.M. (2008): Sources of debris flow material in burned area. Geomorphology, 96, 310–321.

SCHÄLCHLI, U. (1991): Morphologie und Strömungsverhältnisse in Gebirgsbächen: Ein Verfahren zur Festlegung von Restwasserabflüssen. Mitteilungen der Versuchsanstalt für Wasserbau, Hydrologie und Glaziologie, Eidgenössische Technische Hochschule Zürich, Nr. 113, 112p.

SCHEIDEGGER, J. (1970): Theoretical Geomorphology. 2. Auflage, Springer, Berlin.

SCHEIDL, C., RICKENMANN, D. (2010): Empirical prediction of debris-flow mobility and deposition on fans. Earth Surface Processes and Landforms, 35, 157–173.

SCHEIDL, C., RICKENMANN, D. (2011): TopFlowDF – A simple GIS based model to simulate debris-flow runout on the fan. In: R. Genevois, D.L. Hamilton & A. Prestininzi, Debris-Flow Hazards Mitigation: Mechanics, Prediction, and Assessment; Proceedings 5th International DFHM Conference, Padova, Italy, June 14–17, 2011, pp. 253–262. Roma: Casa Editrice Università La Sapienza. doi:10.4408/IJEGE.2011-03.B–030.

SCHEIDL, C., CHIARI, M., KAITNA, R., MÜLLEGGER, M., KRAWTSCHUK, A., ZIMMERMANN, T., PROSKE, D. (2013): Analysing debris-flow impact models, based on a small scale modelling approach. Survey of Geophysics, 34, 121–140.

SCHEUNER T., KEUSEN, H.R., MCARDELL, B.W., HUGGEL, C. (2009): Murgangmodellierung mit dynamisch-physikalischem und GIS-basiertem Fliessmodell. Fallbeispiel Rotlauigraben, Guttannen, August 2005. Wasser, Energie, Luft, 101, 15–21.

SCHILLING, M., HUNZIKER, R. (1995): Programmpaket MORMO – Grundlagen. In: Mathematische Modelle offener Gerinne, ÖWAV-Seminar Konstruktiver Wasserbau – Landschaftswasserbau, 21. Nov. 1995, Bd. 17, pp. 91–104.

SCHMOCKER, L., HAGER, W.H. (2011): Probability of drift blockage at bridge decks. Journal of Hydraulic Engineering, 137, 470–479.

SCHNEIDER, J.M., RICKENMANN, D., TUROWSKI, J.M., BUNTE, K., KIRCHNER, J.W. (2014): Applicability of bedload transport models for mixed-size sediments in steep streams considering macro-roughness. Water Resources Research, 51, 5260–5283.

SCHNEUWLY-BOLLSCHWEILER, M., CORONA, C., STOFFEL, M. (2013): How to improve dating quality and reduce noise in tree-ring based debris-flow reconstructions. Quaternary Geochronology, 18, 110–118.

SCHOKLITSCH, A. (1914): Über Schleppkraft und Geschiebebewegung. Engelmann, Leipzig und Berlin.

SCHREINER, A. (1997): Einführung in die Quartärgeologie. 2. Auflage, E. Schweizerbart'sche Verlagsbuchhandlung, Stuttgart, 257p.

SCHRÖDER, R.C.M. (1994): Technische Hydraulik – Kompendium für den Wasserbau. Springer, Berlin.

SCHÜRCH, P., DENSMORE, A.L., ROSSER, N.J., MCARDELL, B.W. (2011): Dynamic controls on erosion and deposition on debris-flow fans. Geology, 39, 827–830.

SHIELDS, A. (1936): Anwendung der Ähnlichkeitsmechanik und der Turbulenzforschung auf die Geschiebebewegung. Mitteilungen der Preussischen Versuchsanstalt für Wasserbau und Schiffsbau, Berlin, Heft 26.

SINGH, V.P. (2003): On the theories of hydraulic geometry. International Journal of Sediment Research, 18(3), 196–218.

SLAYMAKER, O. (1988): The distinctive attributes of debris torrents. Hydrological Sciences Journal. 33(6), 567–573.

SMART, G.M., JÄGGI, M.N.R. (1983): Sedimenttransport in steilen Gerinnen. Mitteilungen der Versuchsanstalt für Wasserbau, Hydrologie und Glaziologie, Eidgenössische Technische Hochschule Zürich, 64, 188p.

SMART, G.M., DUNCAN, M.J., WALSH, J.M. (2002): Relatively rough flow resistance equations, Journal of Hydraulic Engineering, 128, 568–578.

SPREAFICO, M., LEHMANN, C., NAEF, O. (1996): Empfehlung zur Abschätzung von Feststofffrachten in Wildbächen. Teil I: Handbuch, 46p. + Anhang; Teil II: Fachliche Grundlagen, 113p. Groupe de travail pour l'hydrologie operationelle (GHO), Mitteilung Nr. 4, Landeshydrologie und –geologie, Bern.

SPREAFICO, M., LEHMANN, CH., JAKOB, A., GRASSO, A. (2005): Feststoffbeobachtung in der Schweiz. Berichte des Bundesamtes für Wasser und Geologie, Serie Wasser, Nr. 8, Bern.

STÄHLI, M., SÄTTELE, M., HUGGEL, C., McARDELL, B.W., LEHMANN, P., VAN HERWIJNEN, A., BERNE, A., SCHLEISS, M., FERRARI, A., KOS, A., OR, D., SPRINGMAN, S.M. (2015): Monitoring and prediction in early warning systems for rapid mass movements. Nat. Hazards Earth System Sciences, 15, 4, 905–917.

STINY, J. (1931): Die geologischen Grundlagen der Verbauung der Geschiebeherde in Gewässern. Springer, Wien.

STRICKLER, A. (1923): Beiträge zur Frage der Geschwindigkeitsformel und der Rauhigkeitszahlen für Ströme, Kanäle und geschlossene Leitungen. Sekretariat des Eidg. Amtes für Wasserwirtschaft, Mitteilungen des Amtes für Wasserwirtschaft, Bern, Nr. 16.

SWARTZ, M., McARDELL, B., BARTELT, P. (2003): Interpretation of the August 2000 Schipfenbach debris flow event using numerical models. In: Mitteilungen der Versuchsanstalt für Wasserbau, Hydrologie und Glaziologie, Eidgenössische Technische Hochschule Zürich, Nr. 184, pp. 51–60.

TAKAHASHI, T. (1981): Estimation of potential debris flows and their hazardous zones; soft countermeasures for a disaster. Journal of Natural Disaster Science, 3(1), 57–89.

TAKAHASHI, T. (1987): High velocity flow in steep erodible channels. Proc. XXII IAHR Congress, Lausanne, Switzerland, Technical Session A, pp. 42–53.

TAKAHASHI, T. (1991): Debris Flow. IAHR Monograph Series, Balkema Publishers, the Netherlands.

TAKEI, A. (1984): Interdependence of sediment budget between individual torrents and a river-system. Proc. International Symposium Interpraevent, Villach, Austria, Bd. 2, 35–48.

TECCA, P., GENEVOIS, R., DEGANUTTI, A., ARMENTO, M. (2007): Numerical modelling of two debris flows in the Dolomites (Northeastern Italian Alps). In C.L. Chen & J.J. Major (eds), 4th Int. Conf. on Debris-Flow Hazards Mitigation. Millpress, Rotterdam, The Netherlands, pp. 179–188.

THELER, D., REYNARD, E., LAMBIEL, C., BARDOU, E. (2010): The contribution of geomorphological mapping to sediment transfer evaluation in small alpine catchments. Geomorphology, 124, 113–123.

THOMPSON. S.M., CAMPELL, P.L. (1979): Hydraulics of a large channel paved with boulders. Journal of Hydraulic Research, 17, 341–354.

TOGNACCA, C., BEZZOLA, G.R., MINOR, H.-E. (2000): Threshold criterion for debris-flow initiation due to channel-bed failure. In G.F. Wieczorek & N.D. Naeser (eds), Debris-Flow Hazards Mitigation: Mechanics, Prediction, and Assessment, Proceedings 2nd International DFHM Conference, Taipei, Taiwan, August 16–18, 2000, pp. 89–97. Rotterdam: Balkema.

TOYOS, G., GUNASEKERA, R., ZANCHETTA, G., OPPENHEIMER, C., SULPIZIO, R., FAVALLI, M., PARESCHI, M.T. (2008): GIS-assisted modelling for debris flow hazard assessment based on the events of May 1998 in the area of Sarno, Southern Italy: II. Velocity and dynamic pressure. Earth Surface Processes and Landforms, 33, 1693–1708.

TUROWSKI, J.M., YAGER, E.M., BADOUX, A., RICKENMANN, D., MOLNAR, P. (2009): The impact of exceptional events on erosion, bedload transport and channel stability in a step-pool channel. Earth Surface Processes and Landforms, 34, 1661–1673.

VanDine, D.F. (1985): Debris flows and debris torrents in the Southern Canadian Cordillera. Canadian Geotechnical Journal, 22, 44–68.

VanDine, D.F. (1996): Debris flow control structures for forest engineering. Province of British Columbia, Ministry of Forests Research Program, Working Paper 22/(1996), 75p.

VAW (1992): Murgänge 1987: Dokumentation und Analyse. Unveröffentlichter Bericht, No. 97.6, Versuchsanstalt für Wasserbau, Hydrologie und Glaziologie (VAW), Eidgenössische Technische Hochschule Zürich.

Vetsch, D., Fäh, R., Fahrsi, D., Müller, R. (2005): BASEMENT – Ein objektorientiertes Softwaresystem zur numerischen Simulation von Naturgefahren. In: Mitteilungen der Versuchsanstalt für Wasserbau, Hydrologie und Glaziologie, Eidgenössische Technische Hochschule Zürich, Nr. 190, p. 201–212.

von Ruette, J., Lehmann, P., Or, D. (2013): Rainfall-triggered shallow landslides at catchment scale: Threshold mechanics-based modeling for abruptness and localization. Water Resources Research, 49, 6266–6285.

Waldner, P., Rickli, C., Köchli, D., Usbeck, T., Schmocker, L., Sutter, F. (2007): Schwemmholz. In: G.R. Bezzola & C. Hegg (eds), Ereignisanalyse Hochwasser 2005, Teil 1 – Prozesse, Schäden und erste Einordnung. Bundesamt für Umwelt BAFU, Bern, Eidgenössische Forschungsanstalt für Wald, Schnee und Landschaft WSL, Birmensdorf, pp. 181–193.

Waldner, P., Schmocker, L., Sutter, F., Rickenmann, D., Rickli, C., Lange, D., Köchli, D. (2008): Schwemmholzbilanzen. In: G.R. Bezzola & C. Hegg (eds), Ereignisanalyse Hochwasser 2005: Teil 2 – Analyse von Prozessen, Massnahmen und Gefahrengrundlagen. Bundesamt für Umwelt BAFU, Bern, Eidgenössische Forschungsanstalt WSL, Birmensdorf, pp. 136–143.

Ward, T.J. (1986): Discussion of "Sediment transport formula for steep channels" by G.M. Smart, Journal of Hydraulic Engineering, 112, 989–990.

Warrick, J.A., Rubin, D.M., Ruggiero, P., Harney, J., Draut, A.E., Buscombe, D. (2009): Cobble cam: grain-size measurements of sand to boulder from digital photographs and autocorrelation analyses. Earth Surface Processes and Landforms, 34, 1811–1821.

Webb, B.W., Reid, I., Bathurst, J.C., Carling, P.A., Walling, D.E. (1997): Sediment erosion, transport and deposition. In: C.R. Thorne, R.D. Hey & M.D. Newson (eds), Applied Fluvial Geomorphology for River Engineering and Management, pp. 95–135. Chichester: John Wiley.

Weichert, R., Bezzola, G.R. (2002): Einfluss von Makrorauigkeiten auf die Stabilität alpiner Gewässer. Wasser, Energie, Luft, 94, 259–264.

Whittaker, J.G., Jäggi, M. (1986): Blockschwellen. Mitteilungen der Versuchanstalt für Wasserbau, Hydrologie und Glaziologie, Eidgenössische Technische Hochschule Zürich, Nr. 91, 187p.

Whittaker, J.G., Hickman, W.E., Croad, R.N. (1988): Riverbed Stabilisation with placed blocks. Report 3-88/3, Central Laboratories, Works and Development Services Corporation, Lower Hutt, NZ.

Wichmann, V. Heckmann, T., Haas, F., Becht, M. (2009): A new modelling approach to delineate the spatial extent of alpine sediment cascades. Geomorphology, 111, 70–78.

Wicks, J.M., Bathurst, J.C. (1996): SHESED: a physically based, distributed erosion and sediment yield component for the SHE hydrological modelling system. Journal of Hydrology, 175, 213–238.

Wilcock, P.R., Crowe, J.C. (2003): Surface-based transport model for mixed-size sediment. Journal of Hydraulic Engineering, 129, 120–128.

Wilford, D.J., Sakals, M.E., Innes, J.L., Sidle, R.C., Bergerud, W.A. (2004): Recognition of debris flow, debris flood and flood hazard through watershed morphometrics. Landslides, 1, 61–66.

WOHL, E.E. (2000): Mountain Rivers. Water Resources Monograph, American Geophysical Union, Washington DC, USA, 320p.

WOHL, E., BECKMAN, ND. (2014): Controls on the longitudinal distribution of channel-spanning logjams in the Colorado Front Range, USA. River Research and Applications, 30, 112–131.

WOHL, E., CENDERELLI, D.A., DWIRE, K.A., RYAN-BURKETT, S.E., YOUNG, M.K., FAUSCH, K.D. (2010): Large in-stream wood studies: a call for common metrics. Earth Surf. Process. Landf., 35, 618–625.

WONG, M., PARKER, G. (2006): Reanalysis and correction of bed-load relation of Meyer-Peter and Müller using their own database. Journal of Hydraulic Engineering, 132, 1159–1168.

YAGER, E.M., KIRCHNER, J.W., DIETRICH, W.E. (2007): Calculating bed load transport in steep boulder bed channels. Water Resources Research, 43, W07418. doi:10.1029/2006WR005432.

ZANUTTIGH, B., LAMBERTI, A. (2006): Experimental analysis of the impact of dry avalanches on structures and implication for debris flows. Journal of Hydraulic Research, 44, 522–534.

ZELLER, J. (1963): Einführung in den Sedimenttransport offener Gerinne. Mitteilungen der Versuchsanstalt für Wasserbau und Erdbau, Eidgenössische Technische Hochschule Zürich, Nr. 62, [= Sonderdruck aus Schweizerische Bauzeitung, 81. Jg., Hefte 34, 35, 36].

ZELLER, J. (1985): Feststoffmessung in kleinen Gebirgseinzugsgebieten. Wasser, Energie, Luft, 77, 246–251.

ZELLER, J. (1996): Der Kstr-Koeffizient in der Geschwindigkeitsgleichung von Strickler und dessen Problematik. Proc. International Symposium Interpraevent, Garmisch-Partenkirchen, Germany, Bd. 4, pp. 63–74.

ZELLER, J., TRÜMPLER, J. (1984): Rutschungsentwässerungen – Hinweise zur Bemessung steiler Entwässerungsgräben. Eidgenössische Anstalt für das forstliche Versuchswesen, Birmensdorf, 276p.

ZIMMERMANN, A. (2010): Flow resistance in steep streams: an experimental study. Water Resources Research, 46, W09536. doi:10.1029/2009WR007913.

ZIMMERMANN, M., RICKENMANN, D. (1992): Beurteilung von Murgängen in der Schweiz: Meteorologische Ursachen und charakteristische Parameter zum Ablauf. Proc. International Symposium Interpraevent, Bern, Switzerland, Bd. 2, 153–163.

ZIMMERMANN, M., MANI, P., GAMMA, P., GSTEIGER, P., HEINIGER, O., HUNZIKER, G. (1997a): Murganggefahr und Klimaänderung – ein GIS-basierter Ansatz. Schlussbericht NFP 31, Verlag der Fachvereine, Eidgenössische Technische Hochschule Zürich, Schweiz, 161p.

ZIMMERMANN, M., MANI, P., ROMANG, H. (1997b): Magnitude-frequency aspects of Alpine debris flows. Eclogae Geologicae Helvetiae, 90, 415–420.

ZIMMERMANN, M., LEHMANN, C. (1999): Geschiebefracht in Wildbächen: Grundlagen und Schätzverfahren. Wasser, Energie, Luft, 91, 189–194.

ZOLLINGER, F. (1983): Die Vorgänge in einem Geschiebeablagerungsplatz – Ihre Morphologie und die Möglichkeiten einer Steuerung. Dissertation Nr. 7419, Eidgenössische Technische Hochschule Zürich.

List of symbols

a	coefficient in logarithmic flow resistance law (often $a = 12$)
a_g	coefficient in equation for critical dimensionless unit discharge $q_c{}^*$
a_k	correction factor to account for the slope parallel component of the weight of sediment particles
a_1	coefficient in VPE-equation (mean value $a_1 = 6.5$)
a_2	coefficient in VPE-equation (mean value $a_2 = 2.5$)
A	flow cross-sectional area
A_C	surface area of a hydrologic catchment
A_V	factor (numerator) in equation for the runout distance of a debris flow on the fan
B	width of flow cross-section (at an opening)
d	stem diameter of large wood pieces
d_{Wmax}	maximum dimension of the root part (large wood pieces)
d_{Wmin}	minimum dimension of the root part (large wood pieces)
$d_W{}^*$	$= (d_{Wmax}\, d_{Wmin}\, L_b)^{1/3}$ = mean dimension of the root part (large wood pieces)
D_x	characteristic grain size for which $x\%$ oft the sediment mixture are finer
D_{mi}	mean grain size of the grain size class i
D_{max}	maximum grain size
D_R	duration of a rainfall event
e	exponent in equation for the reduced energy slope
f	$= 8\,(v^*/V)^2$ = friction coefficient according to DARCY-WEISBACH
f_o	DARCY-WEISBACH coefficient for the grain or base-level resistance
f_{add}	DARCY-WEISBACH coefficient for the macro-roughness resistance
f_{tot}	DARCY-WEISBACH coefficient for the total resistance
Fr	$= V/(gh)^{0.5}$ = FROUDE number
g	gravitational acceleration
G	factor (denominator) in equation for the runout distance of a debris flow on the fan
GF	sediment volume (including pore volume of sediment deposits)
h	flow depth
h_u	flow depth in the approach channel (of a debris flow) upstream of the fan apex
H	clear height of an opening of a flow cross-section
H_e	elevation difference between starting point and most distal deposition point of a debris flow

I mean rainfall intensity

K Torrentiality factor (after KRONFELLNER-KRAUS)

k_s equivalente roughness heigth ("sand roughness")

k_{St} STRICKLER coefficient (for total flow resistance)

L total runout distance of a debris flow

L_c length of active channel (regarding erosion during a torrential event)

L_f length of debris-flow deposits on a fan

L_{max} maximum total runout distance of a debris flow

L_b length of trunk extension of large wood pieces

L_W length of log pieces (large woody debris)

M sediment volume of a debris flow (typically estimated from all deposits of an event; includes also pore volume; may refer to single debris-flow surges in case of observations from an automatic monitoring system)

Me MELTON number

n_o MANNING coefficient for the grain or base-level resistance

n_{tot} MANNING coefficient for the total resistance

p_d dynamic impact pressure due to debris flow

p_{Fui} cumulative frequency (as a fraction) of the Fuller distribution for grains with $D \leq D_i$

p_i cumulative frequency (as a fraction) of the grain size distribution for grains with $D \leq D_i$

p_v clogging probability due to large woody debris

Δp_i relative proportion of a grain-size class i of all sediments

P wetted perimeter of a flow cross-section

q unit (water) discharge in a channel (per meter channel width)

q_b bedload transport rate per meter channel width

q_c critical discharge at initiation of bedload motion (per meter channel width)

$q_{c,D}$ critical discharge at breaking up of an armor layer (per meter channel width)

$q_{c,B}$ critical discharge at destabilization of a block ramp (per meter channel width)

q_c^* critical dimensionless discharge (initiation of debris flow, bedload transport)

q^{**} $= q/(gSD_{84}^3)^{0.5} =$ dimensionless discharge (per meter channel width)

Q $= qW =$ total (water) discharge in channel (over entire channel width)

Q_B $= q_bW =$ total bedload transport rate (over entire channel width)

Q_c $= q_cW =$ critical discharge at initiation of bedload motion

$Q_{c,D}$ $= q_{c,D}W =$ critical discharge at breaking up of an armor layer

Q_p maximum discharge (of a debris-flow surge)

Q_S sediment supply from upstream and/or from tributaries

R $= A/P =$ hydraulic radius

s $= \rho_s/\rho =$ ratio of sediment density to water density

S channel slope (or friction slope) (in all equations of this document S has to be used with the units [m/m] und not with the units [%])

S_c mean channel slope upstream of the fan apex

S_f mean channel slope on the fan (or mean fan slope)

S_k $= S\,a_k =$ corrected channel slope to account for the slope parallel component of the weight of sediment particles

S_{red} reduced energy slope (or reduced channel slope) to account for macro-roughness effects in the calculation of bedload transport

S_R friction slope of a debris flow on the fan
T_e maximum erosion depth in a torrent channel
U^{**} $= V/(gSD_{84})^{0.5}$ = dimensionless mean flow velocity
v^* $= (ghS)^{0.5}$ = shear velocity
V mean flow velocity (water or debris flow)
V_o virtual mean flow velocity, related to base-level resistance
V_{tot} (effectiv) mean flow velocity, related to total resistance
V_u flow velocity in the approach channel (of a debris flow) upstream of the fan apex
W width of the channel
y_R height of the lowest roughness layer (in open channel flow)

GREEK SYMBOLS

α_d coefficient for the calculation of the dynamic impact pressure due to debris flow
α_g exponent in equation for the critical dimensionless discharge q_c^*
α_o prefactor in a bedload transport equation
β angle of the channel (or angle of a depositional reach)
β_d angle of impact for the calculation of the dynamic impact pressure due to debris flow
β_u angle of the channel upstream of the fan apex (debris-flow runout calculation)
γ exponent of the "hiding function"
γ_s shear rate (change of flow velocity/change of flow depth)
ε_o roughness height
κ VON KARMAN constant (= 0.4)
μ dynamic viscosity
ρ density of water
ρ_M density of debris-flow mixture (including solids and water)
ρ_s density of solids (sediment particles)
θ $= hS/[(s-1)D]$ = dimensionless bed shear stress
θ' = reduced dimensionless bed shear stress, accounting for energy losses due to form or macro-roughness
θ_c = critical dimensionless bed shear stress at initiation of bedload motion
τ $= \rho ghS$ = bed shear stress
τ_B shear strength, yield stress (Bingham fluid model)
φ_s natural angle of repose of submerged sediment
Φ_b $= q_b/[(s-1)gD^3]^{0.5}$ = dimensionless bedload transport rate (per meter channel width)

Subject index

Printed and bound by CPI Group (UK) Ltd, Croydon, CR0 4YY

24/10/2024

01778286-0012